Multidrug-Resistant Fungi

Multidrug-Resistant Fungi

Guest Editors

Antônio Ferreira-Pereira
Daniel Clemente de Moraes

MDPI

Basel • Beijing • Wuhan • Barcelona • Belgrade • Novi Sad • Cluj • Manchester

Guest Editors

Antônio Ferreira-Pereira
Departamento de
Microbiologia Geral
Universidade Federal do Rio
de Janeiro
Rio de Janeiro
Brazil

Daniel Clemente de Moraes
Departamento de
Microbiologia Geral
Universidade Federal do Rio
de Janeiro (UFRJ)
Rio de Janeiro
Brazil

Editorial Office
MDPI AG
Grosspeteranlage 5
4052 Basel, Switzerland

This is a reprint of the Special Issue, published open access by the journal *Journal of Fungi* (ISSN 2309-608X), freely accessible at: https://www.mdpi.com/journal/jof/special_issues/HY8I5A7A4I.

For citation purposes, cite each article independently as indicated on the article page online and as indicated below:

Lastname, A.A.; Lastname, B.B. Article Title. *Journal Name* **Year**, *Volume Number*, Page Range.

ISBN 978-3-7258-7346-3 (Hbk)
ISBN 978-3-7258-7347-0 (PDF)
https://doi.org/10.3390/books978-3-7258-7347-0

Contents

Journal of *Fungi*

MDPI

Editorial

Multidrug-Resistant Fungi

Daniel Clemente de Moraes and Antônio Ferreira-Pereira *

Laboratório de Bioquímica Microbiana, Departamento de Microbiologia Geral, Instituto de Microbiologia Paulo de Góes, Universidade Federal do Rio de Janeiro, Rio de Janeiro 21941-902, Brazil
* Correspondence: apereira@micro.ufrj.br

Multidrug resistance in fungi is a growing challenge to global public health, resulting in ineffective treatments and thus high mortality rates. This phenomenon is boosted by several factors, such as the indiscriminate use of antifungals, the ability of fungi to evolve resistance mechanisms rapidly, and the dissemination of resistant strains through healthcare facilities and the overall environment. Understanding the molecular basis of resistance and the routes of dissemination of fungi is essential to allow the development of strategies that are able to treat and control resistant fungal infections.

In this Special Issue, eight original articles were published. Yadav et al. (2023) reported the isolation of *Candida auris* from the ears of four dogs in India [1]. All the strains presented high MIC values for fluconazole and were susceptible to other azoles, such as voriconazole and posaconazole, and to drugs belonging to different classes, such as amphotericin B, flucytosine, and echinocandins. According to the World Health Organization (WHO) fungal priority list, *C. auris* is a critical pathogen due to its intrinsic resistance to antifungal agents, the unacceptable mortality rates of its infection, and its ability to spread in hospitals. The study conducted by Yadav et al. described that animals could act as reservoirs for *C. auris*, which may suggest the possibility of interspecies transmission and reinforces the need for additional studies regarding the epidemiology of this fungus.

Casimiro-Ramos et al. (2024) assessed the genome of *C. auris* 20-1498, the first resistant *C. auris* isolated from Mexico [2]. This study detected a point mutation in ERG11, justifying this strain's resistance to fluconazole. *C. auris* 20-1498 displayed thermo-tolerance and halotolerance, features that could facilitate its survival on several surfaces, such as the human body and healthcare facilities.

The study conducted by Parra et al. (2024) evaluated the expression of genes related to azole resistance in *C. glabrata* strains isolated from ICU Colombian patients and observed that exposure to fluconazole can downregulate or upregulate the expression of these genes [3]. In addition, docking analysis showed a low affinity of fluconazole to lanosterol 14-alpha-demethylase, which may explain the high degree of resistance of *C. glabrata* to this antifungal agent.

In order to better understand the adaptation of *Aspergillus fumigatus* to azoles, the study conducted by Hokken et al. (2023) investigated the transcriptional response of three *Aspergillus fumigatus* isolates to azole compounds, such as itraconazole and isavuconazole [4]. Understanding the genes and pathways involved in adaptation to azole compounds could lead to the creation of new antifungal drugs or the improvement of existing therapies, enabling overcoming azole resistance. The azoles tested in this study affected the expression of genes related to sterol metabolism, adhesion, efflux pumps, secondary metabolites, and stress response pathways.

Lucio et al. (2024) assessed the relation between the Mismatch Repair Protein Msh6 and antifungal resistance in *Aspergillus fumigatus* [5]. Deletion of Msh6 led to a higher tendency for developing resistance to non-azole fungicides. The findings suggest that mutations in Msh6 may contribute to the multidrug resistance observed in clinical and environmental isolates of *A. fumigatus*. This knowledge is important for developing new strategies to manage resistant Aspergillus infections.

Three of the studies published in this Special Issue assessed the effect of different substances against resistant fungi. Using two *C. albicans* and one *C. glabrata* clinical isolates, Rollin-Pinheiro et al. (2023) verified the antifungal activity of the depsipeptide aureobasidin A [6]. This compound affected Candida planktonic cells and biofilms. It led to oxidative stress, diminished chitin content, and disturbed mitochondria, among other effects within the cells. Notably, aureobasidin A inhibited CaCdr2p and CaMdr1p in *Saccharomyces cerevisiae* mutant models. These are two of the three main efflux transporters related to the multidrug resistance phenotype in *C. albicans*. The effect of this inhibition was seen in an in vivo model, where the combination of aureobasidin A with fluconazole increased the survival of *Caenorhabditis elegans*.

Silva et al. (2023) developed a new synthetic route to produce Altissimacoumarin D. They tested the ability of this molecule and analogues to inhibit *Candida albicans* efflux pumps (CaCdr1p, CaCdr2p, and CaMdr1p), using *S. cerevisiae*-and *C. albicans*-resistant strains [7]. Two compounds (ACS47 and ACS50) inhibited ABC and MFS transporters, hence reversing the multidrug resistance phenotype. Moreover, they did not display toxicity in vitro and in silico. Therefore, these coumarins may be potential candidates for use in association with fluconazole to treat *C. albicans* azole-resistant infections.

In a drug repurposing study, Xisto et al. (2023) assessed the effect of miltefosine on Mucorales fungi, such as *Rhizopus oryzae* and *Mucor velutinosus* [8]. Miltefosine presented fungistatic activity, disrupted biofilms, and affected the cell wall content. Moreover, it likely interacts with fungal lipids. This study is a great example of the importance of drug repurposing to discover new antimicrobial drugs.

This Special Issue presents manuscripts that are collectively crucial to the field of antifungal resistance, covering the dissemination of resistant fungi, mechanisms of resistance, and the discovery of potential antifungal agents. We are grateful to the authors for their contributions to this Special Issue and to the study of resistance in fungi. These microorganisms are severely neglected, and it is through high-quality research that we can hope to overcome resistant fungal infections. We also extend our gratitude to the reviewers for their valuable corrections and suggestions.

Conflicts of Interest: The authors declare no conflicts of interest.

References

1. Yadav, A.; Wang, Y.; Jain, K.; Panwar, V.A.R.; Kaur, H.; Kasana, V.; Xu, J.; Chowdhary, A. *Candida auris* in Dog Ears. *J. Fungi* **2023**, *9*, 720. [CrossRef] [PubMed]
2. Casimiro-Ramos, A.; Bautista-Crescencio, C.; Vidal-Montiel, A.; González, G.M.; Hernández-García, J.A.; Hernández-Rodríguez, C.; Villa-Tanaca, L. Comparative Genomics of the First Resistant *Candida auris* Strain Isolated in Mexico: Phylogenomic and Pan-Genomic Analysis and Mutations Associated with Antifungal Resistance. *J. Fungi* **2024**, *10*, 392. [CrossRef] [PubMed]
3. Cárdenas Parra, L.Y.; Rojas Rodríguez, A.E.; Pérez Cárdenas, J.E.; Pérez-Agudelo, J.M. Molecular Evaluation of the MRNA Expression of the ERG11, ERG3, CgCDR1, and CgSNQ2 Genes Linked to Fluconazole Resistance in *Candida glabrata* in a Colombian Population. *J. Fungi* **2024**, *10*, 509. [CrossRef] [PubMed]
4. Hokken, M.W.J.; Coolen, J.P.M.; Steenbreker, H.; Zoll, J.; Baltussen, T.J.H.; Verweij, P.E.; Melchers, W.J.G. The Transcriptome Response to Azole Compounds in *Aspergillus fumigatus* Shows Differential Gene Expression across Pathways Essential for Azole Resistance and Cell Survival. *J. Fungi* **2023**, *9*, 807. [CrossRef]
5. Lucio, J.; Gonzalez-Jimenez, I.; Roldan, A.; Amich, J.; Alcazar-Fuoli, L.; Mellado, E. Importance of the *Aspergillus fumigatus* Mismatch Repair Protein Msh6 in Antifungal Resistance Development. *J. Fungi* **2024**, *10*, 210. [CrossRef] [PubMed]
6. Rollin-Pinheiro, R.; de Moraes, D.C.; Bayona-Pacheco, B.; Curvelo, J.A.R.; dos Santos-Freitas, G.M.P.; Xisto, M.I.D.d.S.; Borba-Santos, L.P.; Rozental, S.; Ferreira-Pereira, A.; Barreto-Bergter, E. Structural and Functional Alterations Caused by Aureobasidin A in Clinical Resistant Strains of *Candida* spp. *J. Fungi* **2023**, *9*, 1115. [CrossRef] [PubMed]

7. Silva, A.C.; de Moraes, D.C.; do Carmo, D.C.; Gomes, G.C.C.; Ganesan, A.; Lopes, R.S.C.; Ferreira-Pereira, A.; Lopes, C.C. Synthesis of Altissimacoumarin D and Other Prenylated Coumarins and Their Ability to Reverse the Multidrug Resistance Phenotype in *Candida albicans*. *J. Fungi* **2023**, *9*, 758. [CrossRef] [PubMed]
8. Xisto, M.I.D.d.S.; Rollin-Pinheiro, R.; Rochetti, V.P.; de Castro-Almeida, Y.; Borba-Santos, L.P.; dos Santos-Freitas, G.M.P.; Cypriano, J.; Abreu, F.d.Á.; Rozental, S.; Barreto-Bergter, E. Miltefosine: A Repurposing Drug against Mucorales Pathogens. *J. Fungi* **2023**, *9*, 1166. [CrossRef] [PubMed]

Journal of
Fungi

MDPI

Article

Candida auris in Dog Ears

Anamika Yadav [1,2,†], Yue Wang [3,†], Kusum Jain [1,2], Vijay Amrit Raj Panwar [4], Hardeep Kaur [2], Vikas Kasana [1], Jianping Xu [3] and Anuradha Chowdhary [1,*]

1 Medical Mycology Unit, Department of Microbiology, Vallabhbhai Patel Chest Institute, University of Delhi, Delhi 110007, India; yadavanamika761995@gmail.com (A.Y.); kusumjain93@gmail.com (K.J.); vikaskasana532@gmail.com (V.K.)
2 Department of Zoology, Ramjas College, University of Delhi, Delhi 110007, India; hkaur53d@gmail.com
3 Department of Biology, McMaster University, Hamilton, ON L8S 4K1, Canada; wangy660@mcmaster.ca (Y.W.); jpxu@mcmaster.ca (J.X.)
4 In-Patient Department, Sanjay Gandhi Animal Care Centre, New Delhi 110027, India; varspanwar@gmail.com
* Correspondence: chowdhary.anuradha@gmail.com
† These authors contributed equally to this work.

Abstract: *Candida auris* is an emerging global public health threat and is resistant to most antifungal agents. Though fungi are significant pathogens for animals, the role of *C. auris* in animal health remains unexplored. Here, we analysed the microbial cultures of skin and ear swabs of 87 dogs in Delhi and performed fungal meta-barcode sequencing of ear and skin samples of 7 dogs with confirmed otitis externa (OE). Overall, 4.5% of dogs (4/87) with chronic skin infections contained evidence of *C. auris* in their ear canal ($n = 3$) and on their skin surface ($n = 1$). Of the three OE dogs with *C. auris* infection/colonisation, a diversity of fungi was observed, and their meta-barcode ITS sequence reads for *C. auris* ranged from 0.06% to 0.67%. Whole-genome sequencing of six *C. auris* strains obtained in culture from two dogs showed relatedness with Clade I clinical strains. The report highlights the isolation of *C. auris* from an animal source; however, the routes of transmission of this yeast to dogs and the clinical significance of transmission between dogs and humans remain to be investigated.

Keywords: otitis externa; meta-barcode sequencing; ITS; *Candida auris*; dogs; whole-genome sequencing

1. Introduction

Candida auris is a newly emerged human pathogenic yeast and is resistant to most antifungal agents. It has caused a diversity of infections and outbreaks in hospitals and presents a huge treatment challenge for clinicians and public health authorities. First reported in Japan in 2009, *C. auris* has spread rapidly worldwide [1–3] and caused the World Health Organisation to declare it as one of the four 'critical priority' fungal pathogens [4]. Outside hospital settings, *C. auris* has been isolated from the surface of stored apples, tidal marshes, hypersaline environments, and recently from wastewater, suggesting that this yeast can survive in harsh conditions [5–8]. Interestingly, trace sequences identical or highly similar to that of *C. auris* at the fungal barcode locus, i.e., the internal transcribed spacer (ITS) region, were found in metabarcoding data on an ear sample from a Spanish dog with otitis externa (OE) and from the skin of newts in the United Kingdom [9]. Though fungi are significant pathogens for animals, no live culture of *C. auris* has been isolated from animals, and its role in animal health remains to be explored [10]. We hereby report the isolation of *C. auris* from ear samples of two dogs. Additionally, culture-independent assessment of mycobiome analysis (DNA metabarcoding data) showed the presence of *C. auris*-specific ITS reads from the ear and skin of dogs with chronic skin infections.

2. Material and Methods

Swab collection and processing: In this study, we analysed the microbial communities of skin and ear swab samples in culture from a heterogeneous cohort of 87 dogs from a public referral hospital cum shelter in Delhi, India. The skin and ear swabs were collected and processed for bacterial and fungal cultures using the routine diagnostic protocol for suspected skin and ear infections. An ethics review was not requested by the Animal Care Centre because it was considered as a clinical observation study. The ear canal was sampled using sterile swabs (HIMEDIA, Mumbai, India) by rubbing the skin between the vertical–horizontal junctions. The skin swabs were obtained from the presenting lesions covering the dorsal–ventral skin, paws, and groin region. For microbiologic culture, the swabs were streaked directly on standard bacterial culture media, i.e., sheep blood agar (SBA) and MacConkey agar at 37 °C for 24 h. For fungal growth, Sabouraud dextrose agar (SDA) and CHROMagar (Becton, Dickinson, Baltimore, MD, USA) media were used and incubated at 37 °C for 24 h and 48 h, respectively. Additionally, for isolation of *Candida auris*, each swab was inoculated into yeast nitrogen base (YNB) broth with 10% NaCl at 37 °C for 4 days followed by streaking onto the SDA plates [7,11]. All microbial colonies, including bacteria and yeasts on blood agar, MacConkey agar, SDA, and CHROMagar, were identified by matrix-assisted laser desorption ionization–time of flight mass spectrometry (MALDI-TOF MS; Bruker Biotyper OC version 3.1, Daltonics, Bremen, Germany) with a score $\geq$ 2. *Candida auris* colonies obtained in culture were further confirmed by sequencing of internal transcribed spacer region (ITS) within the ribosomal RNA gene cluster [12].

Fungal DNA extraction and ITS amplicon sequencing (Metagenomics): In addition to samples for microbiological cultures, ITS meta-barcode sequencing was conducted for swab samples (skin, n = 2; ear swabs, n = 7) from seven dogs to study the fungal communities in these dogs with OE. Swabs were collected and placed in 0.05 M phosphate buffer for 30 min with a continuous shaking at 200 rpm. Swabs were discarded, and the solution was stored at −20 °C until subsequent analyses [7]. Microbial genomic DNA was extracted from each sample using a column-based QIAamp DNA minikit (Qiagen, Hilden, Germany) and quantified by QUBIT 4 fluorometer (ThermoFisher, USA) using DNA HS assay kit (ThermoFisher). Primers ITS-1, 5′-TCCGTAGGTGAACCTTGCGG-3′, ITS2, 5′-GCTGCGTTCTTCATCGATGA-3′, ITS3, 5′-GCATCGATGAAGAACGCAGC-3′, and ITS4, 5′-TCCTCCGCTTATTGATATGC-3 were used for amplifying the ITS1 and ITS2 regions, respectively. Illumina sequencing platform (NOVASEQ 6000, PIPITS v2.7) was used to perform ITS fungal sequencing, and PIPITS v2.7 was used for analysis [13]. Magic-BLAST was used to map the raw metabarcoding sequences of each sample against the fungal ITS database containing 15,695 fungal species [14]. The reference ITS sequences can be downloaded at https://ftp.ncbi.nlm.nih.gov/refseq/TargetedLoci/Fungi/. To calculate the relative read abundance at the species level in each sample, the read counts of each species were divided by the total reads mapping to the fungal ITS database. Then, the read abundance was pooled into the genera level, and the average read abundance of the samples for each genus was used to select the taxa for the heatmap. A total of 28 genera with average read abundance over 0.11% were included, and the 28 genera covered an average of 98% reads that were mapped to the database. A heatmap was generated using matplotlib and seaborn [15,16]. Additionally, in each sample, we specifically searched for ITS sequences of *C. auris*. For samples that contained *C. auris* ITS sequences (>94% sequence identity over the cleaned and filtered ITS reads), the relative frequencies of *C. auris* ITS sequences in each sample were calculated. To better reveal the differences among samples, the read abundance for each genus was standardized. The standardization was carried out for each taxon by subtracting the minimum and dividing each by its maximum.

Antifungal susceptibility testing: The susceptibility testing of yeasts was performed using the CLSI broth microdilution method, following M27-A3 [17]. The antifungals tested were fluconazole (FLU, Sigma, St. Louis, MO, USA), itraconazole (ITC, Lee Pharma, Hyderabad, India), voriconazole (VRC, Pfizer, Groton, CT, USA), posaconazole (POS, Merck, Whitehouse Station, NJ, USA), isavuconazole (ISA, Basilea Pharmaceutical, Basel,

Switzerland), 5-flucytosine (5-FC, Sigma, St. Louis, MO, USA), micafungin (MFG, Astellas, Toyama, Japan), anidulafungin (AFG, Pfizer, Groton, CT, USA), and amphotericin B (AMB, Sigma, Groton, CT, USA). The drugs were tested for 10 (two-fold) dilutions, and the drug concentration ranges were as follows: FLU, 0.25–>128 mg/L; ITC, VRC, and AMB, 0.03–16 mg/L; POS, ISA, AFG, MFG, 0.015–8 mg/L; 5-FC, 0.125–64 mg/L [18]. *Candida krusei* strain ATCC6258 and *C. parapsilosis* strain ATCC22019 were used as quality controls. All statistical parameters were calculated using Prism version 6.00 (GraphPad Software).

Antibiotic susceptibility testing was conducted for all bacterial isolates using disk diffusion method following CLSI M100, Ed-32. For Gram-positive bacteria, including *Staphylococcus pseudointermedius*, *S. intermedius*, *S. simulans*, and *S. schleiferi*, eight drugs were tested, namely penicillin, ampicillin, clindamycin, vancomycin, azithromycin, linezolid, cefoxitin, and cotrimoxazole. For Gram-negative bacteria, i.e., *Acinetobacter baumanii*, and *Pseudomonas aeruginosa*, ciprofloxacin, levofloxacin, gentamicin, amikacin, meropenem, cefepime, ceftazidime, ceftazidime-clavulanate, piperacillin, and piperacillin-tazobactam were tested. The inhibitory zone diameters obtained around the antibiotic discs were measured after incubation for 24 h at 37 °C and evaluated [19].

Whole Genome Sequencing (WGS): Genomic DNA of all *C. auris* isolates (n = 6 obtained from ear swabs of two dogs with OE) were extracted using a column-based QIAamp DNA minikit (Qiagen, Hilden, Germany) and quantified by a QUBIT 3 Fluorometer using dS DNA HS Dye. WGS libraries were prepared using NEBNext ultra II DNA FS kit (New England Biolabs, Ipswich, MA, USA), as described previously, and sequenced on Illumina HiSeq 4000 [5,11]. For phylogenetic analysis to identify the clade affinity of our strains, one reference genome from each of the four major clades was included for comparison, including strains B8441 (clade I), B11220 (clade II), B11221 (clade III), and B11245 (clade IV). All strains were subjected to NASP pipeline for genome sequencing analysis, as described previously [7]. The phylogenetic tree was constructed using RAxML v8.0.25 under ASC_GTRCAT nucleotide substitution model and 1000 bootstrap replicates [20]. A maximum-likelihood tree was constructed based on SNPs present in at least one sample. This analysis revealed that all of our strains belonged to clade I (see Results below). We then proceeded to conduct an additional phylogenetic analysis using B8441 as reference and including 569 previously reported clade I *C. auris* samples from different countries, including Indian clinical strains [21], strains recovered from apples [7], and strains from the marine environment of the Andaman Islands [5]. In this broad clade I-specific analysis, SNP sites with ambiguous calls in over 0.5% of the samples were removed, and the remaining SNPs were concatenated for each sample. Maximum-likelihood phylogeny was constructed using RAxML-HPC2 on XSEDE in the CIPRES Science Gateway [22]. The tree uses the ASC_GTRCAT nucleotide substitution model and 1000 bootstrap iterations and was visualized with iTOL [23].

3. Results

Population description: Among 87 dogs, 52 (60%) were stray dogs who were admitted in the intensive care unit (ICU) with severe skin lesions due to chronic skin diseases (group IA and IB). In group IA dogs with skin infections, eight had otitis externa with clinical signs of ear infection, i.e., erythema, oedema, erosion, and exudate in the affected ear (Table 1). In group IB all of the remaining 44 dogs exhibited skin infections. The group II composed of 35 pet dogs (40%) who were attending the outpatient services for minor ailments of the gastrointestinal and urinary tracts.

Fungal and bacterial culture: Overall, the ear swabs of all 52 dogs (group IA and IB) yielded microbial growth. On the contrary, the skin samples of only 31% (n = 16) of the dogs had microbial growth, including the isolation of bacteria (n = 8) and yeast (n = 8) in eight dogs each. Furthermore, ear samples and skin samples of 14% (n = 7) and 4% (n = 2) of dogs contained both living yeasts and living bacteria. In the group IA and IB cohorts, *Candida tropicalis* (n = 3) and *Malassezia pachydermatis* (n = 3) were the predominant fungi on the skin (6% each), followed by *Candida glabrata* in a single dog. Also, in two (4%) dogs,

Acinetobacter baumanii and *Staphylococcus pseudointermedius* were detected on the skin. In ear samples, *M. pachydermatis* was the predominant fungal species (*n* = 11, 21% dogs), followed by *C. tropicalis* (*n* = 8) in 15% cases (Table 1).

Table 1. Distribution of yeasts and bacteria on the skin and in the ear of 87 dogs collected from North Delhi, India.

Disease Categories in Dogs (Numbers)	Period of Sampling	Yeasts		Bacterial Isolation (Body Site, No. of Dogs Positive for the Given Species)
		Culture (Body Site, No. of Dogs Positive for the Given Species)	Amplicon Sequencing (ITS Reads for *C. auris*/Total Reads Matched to Fungal ITS; Percentage of ITS Reads)	
Group IA Chronic skin disease with otitis externa (*n* = 8),	27-09-22 to 17-04-23	*C. auris* (E *, *n* = 1) (D1) @ *C. rugosa* (E, *n* = 2), *C. krusei* (E, *n* = 3) (D2) @ *C. lusitaniae* (E, *n* = 1) *C. glabrata* (S #, *n* = 1) (D3) @	*C. auris* from ear swab of D1 dog (34394/27920633; 0.12%) *C. auris* from ear swab of D2 dog (39577/59853350; 0.06%) *C. auris* from skin swab of D3 dog (3231/479708; 0.67%)	*A. baumanii* (E, *n* = 1) *S. pseudointermedius* (S, *n* = 1)
Group IB Chronic skin disease (*n* = 44),	14-08-22 to 17-04-23	*C. auris* (E, *n* = 1) (D4) @ *M. pachydermatis* (E, *n* = 11; S, *n* = 3) *C. tropicalis* (E, *n* = 8; S, *n* = 3), *T. asahii* (E, *n* = 7), *C. glabrata* (E, *n* = 6, S; *n* = 1), *C. rugosa* (E, *n* =1), *C. albicans* (E, *n* = 1),	ND	*A. baumanii* (E, *n* =15; S, *n* = 2) *S. pseudointermedius* (E, *n* = 9; S, *n* = 2) *S. intermedius*, (E, *n* = 1; S, *n* = 1) *K. pneumoniae* (E, *n* = 1) *S. simulans* (S, *n* = 1) *S. schleiferi* (S, *n* = 1) *P. aeruginosa* (E, *n* = 1)
Group II Negative for skin and ear infections (*n* = 35),	11-05-22 to 02-09-22	*C. tropicalis* (E, *n* = 3; S, *n* = 3) *M. pachydermatis* (E, *n* = 3; S, *n* = 1) *C. krusei* (E, *n* = 2) *C. lusitaniae* (S, *n* = 2) *C. parapsilosis* (E, *n* = 1)	ND	*A. baumanii* (E, *n* = 1; S, *n* = 2), *S. pseudointermedius* (E, *n* = 6), *P. aeruginosa* (E, *n* = 3; S, *n* = 1) *S. schleiferi* (E, *n* = 1; S, *n* = 1) *S. intermedius* (S, *n* = 2), *K. pneumoniae* (S, *n* = 1), *E. faecalis* (S, *n* = 1), *B. cereus* (E, *n* = 1)

* ear, @ dogs positive for *C. auris* either by culture or ITS amplicon sequencing, # skin.

In group IA, all cases had confirmed ear infections, and the ear swabs yielded yeasts as the sole etiologic agent in four of eight cases. The ear swabs of the remaining four dogs had both yeasts and bacterial colonies in culture. Notably, a single dog with OE (D1 dog) showed colonisation with multiple yeasts in the ear skin, including *C. auris* (645-V-22), *Candida lusitaniae,* and *Candida rugosa,* along with *Acinetobacter baumanii.* The *C. auris*-positive dog (D1) was admitted with chronic skin infection and OE in the hospital for more than a month before our sampling, and the dog was on topical treatment with miconazole, mupirocin, and ofloxacin. A repeat sampling of the ear canal of this dog after four weeks again yielded *C. auris* (645-V-22_2). Furthermore, the ear swab obtained from another dog (D4) in group IB (with only skin infection) who had no symptoms of otitis was also positive for *C. auris*, yielding five colonies of *C. auris* (D_953_1 to D_953_5) in culture.

Interestingly, 63% (*n* = 22) of the group II cohort (*n* = 35) with no skin or ear infections showed microbial growth from ear and/or skin swabs, but none of the colonies were *C. auris*. Also, a high rate of skin colonisation (43%, *n* = 15) was noted in these otherwise healthy dogs, whereas ear samples of only seven dogs showed microbial growth. Of these seven dogs, both *Candida tropicalis* (*n* = 3) and *M. pachydermatis* (*n* = 3) were observed to be equally distributed in the ear samples, followed by *Candida krusei* (*n* = 2) and *Candida parapsilosis* (*n* = 1). The skin surface was predominately colonised by *C. tropicalis* (*n* = 3, 14%) and *C. lusitaniae* (*n* = 2, 9%), and bacterial colonisation was due to *A. baumanii* (*n* = 2, 5.7%) and *S. intermedius* (*n* = 2, 5.7%).

We investigated healthcare workers' (HCW, $n = 6$) hand swabs for microbial growth. The ICU board of the Animal Facility guided the collection of hand swab cultures from healthcare workers ($n = 6$) handling the dogs and the environmental sampling of the ICU for microbial cultures. Also, the hand swab of a single HCW in the ICU was analysed for fungal diversity using metabarcoding. Hand colonisation by microbes was noted in five of the six HCWs. Overall, a varied spectrum of yeasts, including *C. krusei*, *C. parapsilosis*, *C. tropicalis*, and *Trichosporan asahii*, was observed on the skin surfaces of the hands of HCWs in culture. Additionally, the doors, benches, and floor of the ICU were found to be colonised by *C. tropicalis* and *C. krusei*.

Metagenomics analysis: According to metabarcoding data, evidence of *C. auris* was found in the ear skin samples of two dogs (both with OE) and on the skin surface of a single dog with OE (in the group IA cohort). A total of 248,515,598 ITS reads were obtained from ten samples (nine samples from dogs and one HCW's hands), with an average of 24,851,560 reads per sample. After sequence alignment, 151,552,302 ITS reads were matched to a fungal ITS database with an average of 15,155,230 reads per sample. The length of the high-quality reads was 152 base pairs. The meta-barcode data from the ear of the dog with a positive culture for *C. auris* (D1) were also found positive for 34,394 ITS reads (0.12%) for *C. auris*. In addition, two dogs with negative culture for *C. auris* also showed the presence of 39,577 (0.06%) and 3231 (0.6%) ITS reads for *C. auris* from the ear (D2) and skin (D3), respectively. The heatmap shows the percentage of ITS read abundance of fungi in the 10 samples (Figure 1). The percent ITS reads for all fungal species are given in Table S1.

Interestingly, among the three dogs containing ITS sequences for *C. auris*, there seemed to be a negative correlation between the ITS read abundance for *Malassezia* species and that of *C. auris* (Table S1). In addition, a single ear sample with the highest percentage (52.2%) of ITS reads for *Malassezia* did not have traces of *C. auris*. However, there was a big variation in the frequencies of *Malassezia* ITS sequences (0–52.2%) among *C. auris*-negative dogs. Aside from *C. auris*, other *Candida* species were frequently recovered. Of the seven dogs with OE whose ear swabs were subjected to ITS amplicon sequencing, there was an average of 15,155,230 ITS reads per samples for *Candida* species. In addition, the presence of *C. tropicalis*, *C. krusei*, *C. parapsilosis* and *T. asahii*, in the microbial cultures of the hand swab of an HCW showed correlation with the presence of ITS reads (3.2%) of *Candida* species.

Antifungal susceptibility testing (AFST): The MIC data of the antifungals tested are summarised in Table S2. The two *C. auris* strains (645-V-22, 645-V-22_2) from the D1 dog, were found to be resistant to fluconazole (FLU) (MIC > 128 mg/L), while all of the other five strains isolated from the D4 case showed 2–4-fold lower MICs for FLU (32–64 mg/L). All *C. auris* strains were susceptible to the other tested drugs (Table S2).

All of tested isolates of *Staphylococcus*, *Pseudomonas*, and *Acinetobacter* genera were susceptible to all drugs except *S. pseudointermedius*, which was found to be resistant to clindamycin and azithromycin.

Whole genome sequencing: WGS analysis revealed that two *C. auris* strains (645-V-22, 645-V-22_2) obtained from the D1 dog four weeks apart differed by four SNPs from each other, suggesting that the same strain persisted during the four weeks in the colonised ear. In the D4 dog, four of the five isolates were sequenced, and clonal clustering of all four strains of *C. auris* was observed with a 2–6 SNPs difference among the strains. Notably, these four strains (D_953_1, D_953_2, D_953_3, and D_953_5) showed a difference of 955–965 SNPs from *C. auris* strains isolated from the D1 dog, suggesting the circulation of genetically distinct *C. auris* strains. All four *C. auris* strains from the D4 dog clustered with B8441, a clade I reference strain (Pakistan). Interestingly, the *C. auris* strains from the D4 dog showed close genetic relatedness (8–11 SNPs difference) with the Pakistan clade I reference strain. Phylogenetic analysis of 569 global clade I *C. auris* strains grouped the two D1 strains with two *C. auris* strains causing candidemia (7–14 SNPs) in Delhi, India (Figure 2) [3,21]. These two strains from the D1 dog showed K143R amino acid substitution in the *ERG11* gene and A640V amino acid substitution in the *TAC1B* gene. Notably, four *C.*

auris strains from the D4 dog showed the absence of mutation in azole-resistance-related genes, i.e., *ERG11* and *TAC1B*.

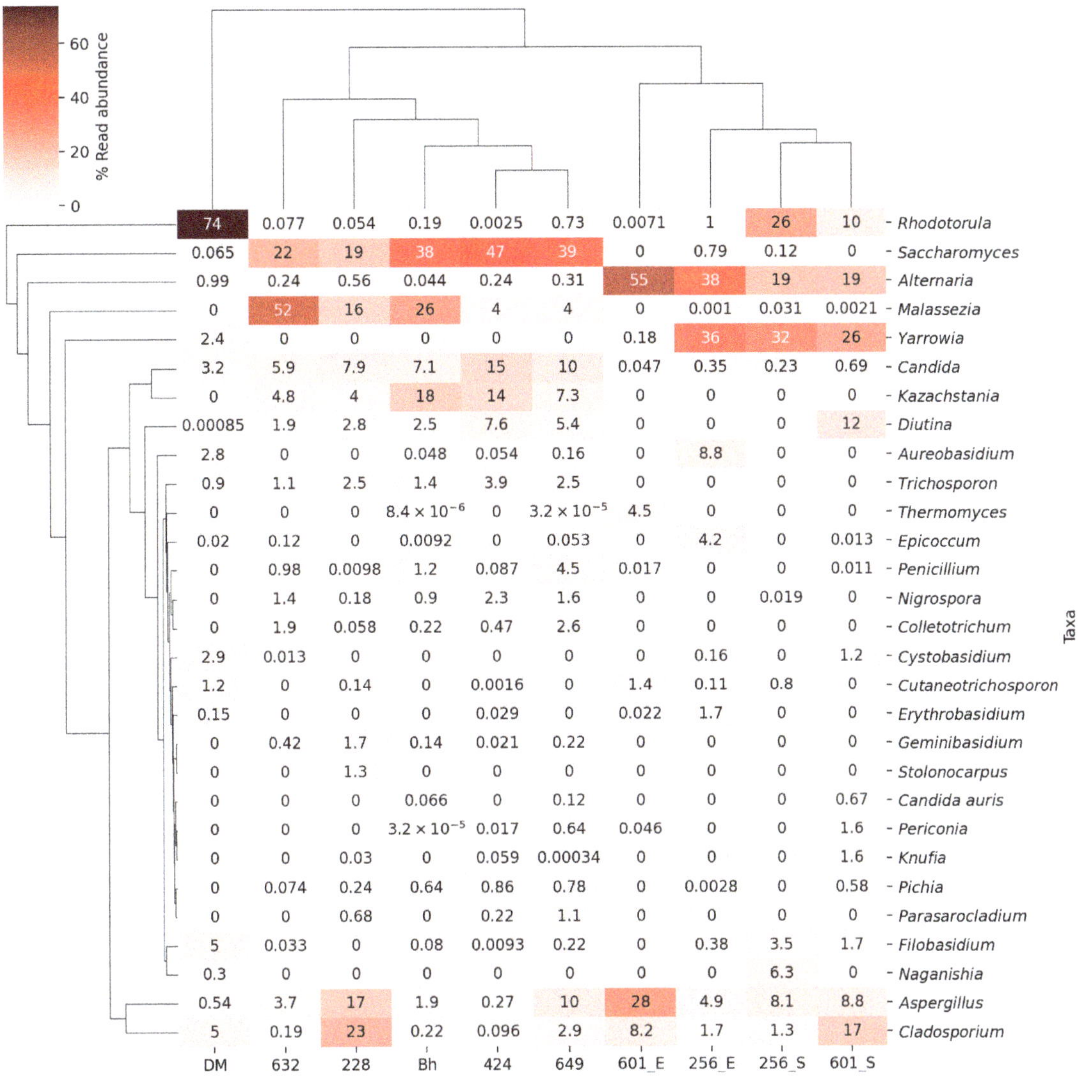

Figure 1. Heatmap showing the distribution of the 28 most common fungal genera in ear and skin samples from seven dogs and one healthcare worker. Different intensities of colours represent the percent read abundance of a genus in each sample. The dendrogram on the left is to indicate the similarity between different genera based on their read abundance profiles. Genera that are closer together on the dendrogram share more similar abundance patterns in our samples. The dendrogram on the top represents the clustering and relationships between the different samples based on their read abundance profiles of listed genera. Samples that are closer together on the dendrogram share more similar read abundance profiles.

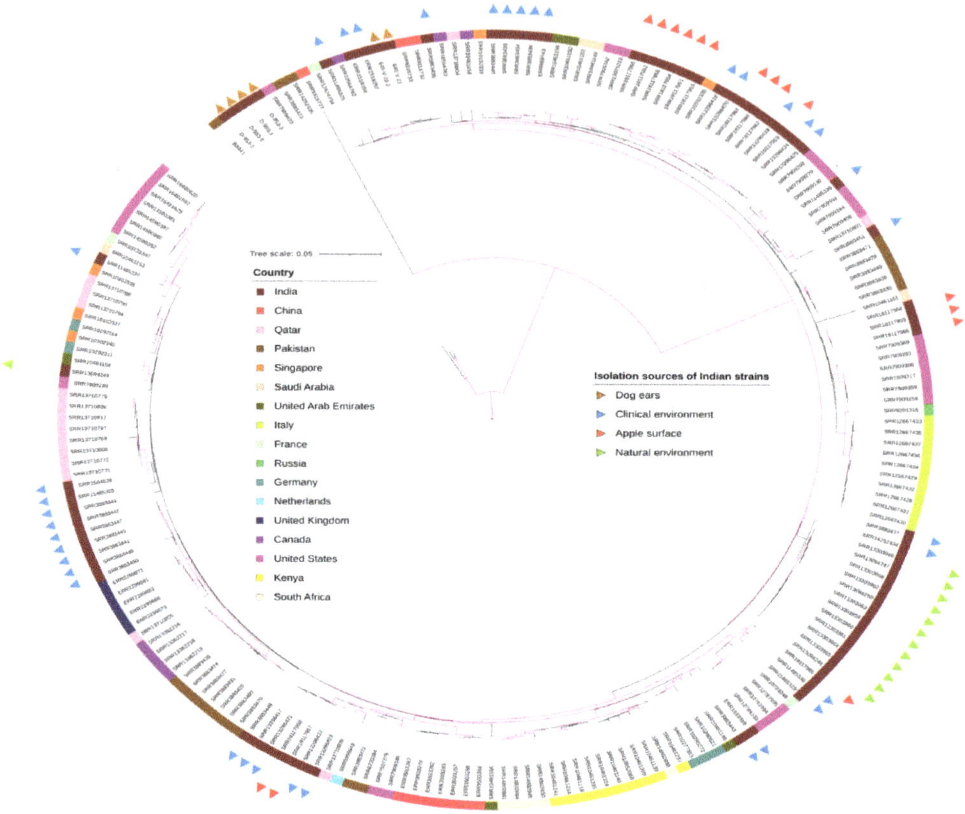

Figure 2. Genetic relationships of *C. auris* isolates from dog ears and clade I strains from other countries inferred based on concatenated genome-wide SNPs. Branches with bootstrap support over 0.95 are highlighted in purple. Indian strains were labelled with triangles in various colours to indicate their isolation sources. The inner colour strips specify the geographic location of the isolates.

4. Discussion

Our report documents for the first time the isolation of live *C. auris* culture from an animal source. Overall, 4 of the 87 dogs (4.5%) contained evidence of *C. auris* infection or colonisation in their ear and on the surface of their skin. Notably, three of the four *C. auris*-positive dogs, either by culture (n = 2) or by ITS meta-barcode sequencing (n = 2), harboured this yeast in the ear canal. Furthermore, among the eight dogs in the OE cohort, two had *C. auris,* suggesting an association of *C. auris* with external ear canal infections.

Candida auris has been previously isolated from ear infections in humans from Korea, Japan, and Iran [1,24,25], but those strains mainly belonged to Clade II and Clade V [3,25,26]. In a recent study from Spain investigating the bacterial and fungal microbiota in the ear canals of dogs with OE, 12 (0.01%) *C. auris*-specific ITS reads out of a total of 81,213 ITS reads were found in a single dog [9,27]. In contrast, in the present study, an ample number of *C. auris* reads accounting for 0.06% and 0.12% was observed in the ear samples, much higher than that in the ear sample of the Spanish dog. In addition, the skin of a single dog was also found to be colonised by *C. auris*, with 3231 (0.67%) ITS reads. The genomic data of *C. auris* strains from the ears of two dogs showed that genetically distinct strains of Clade

I were infecting or colonising the dogs. A close association of *C. auris* strains from a single dog (D4) with the Pakistan clade I reference strain suggested a clonal transmission of this yeast in the Indian subcontinent. Interestingly, this strain with a MIC value of 32 mg/L for fluconazole had no known mutations in genes related to azole resistance, i.e., in the *ERG11* and *TAC1B* genes. In contrast, *C. auris* strains from the D1 dog showed similarity with the Indian clinical strains and exhibited previously known K143R amino acid substitution in the *ERG11* gene and A640V amino acid substitution in the *TAC1B* gene, likely contributing to the high MIC of fluconazole (MIC > 128 mg/L).

It is important to emphasise that in the present study, *C. auris* was not detected on the healthy intact skin of pet dogs using the culture method. However, the ITS amplicon sequencing was not performed on the skin and ear swabs of otherwise healthy dogs. *C. auris* colonised the breached skin of the ears of dogs with chronic skin conditions admitted to the ICU, but colonisation of the fomites samples in the ICU was not observed. This scenario is unlike the colonisation by *C. auris* in humans, where the close inanimate environment is readily contaminated with this yeast due to shedding of skin scales. The fact that, in the present study, the skin of the ears of dogs was found to be colonised by *C. auris* rather than the exposed (ventral and dorsal) skin probably contributed to reduced shedding in the environment. It is worth emphasising the circulation of closely related *C. auris* clade I strain in humans and dogs. However, the prevalence of *C. auris* transmissions between humans and domesticated animals warrants further investigations. Finally, the association of *C. auris* with ears in humans and dogs hints toward the ear canal possibly being an important reservoir of this yeast.

Supplementary Materials: The following supporting information can be downloaded at: https://www.mdpi.com/article/10.3390/jof9070720/s1, Table S1: Distribution of the 28 most common fungal genera in the ten samples (9 samples from dogs and one HCW hands). Individual sheets list the percent ITS reads data for all fungal genera from ten samples. Table S2: In vitro antifungal susceptibility profile of 10 yeast species isolated from dogs (n = 87) against ten antifungal drugs using CLSI-BMD method.

Author Contributions: Conceptualization, A.C.; Data curation, Y.W.; Formal analysis, Y.W., J.X. and A.C.; Investigation, A.Y. and V.A.R.P.; Methodology, A.Y., K.J. and V.K.; Supervision, A.C.; Writing—original draft, A.Y., K.J. and A.C.; Writing—review and editing, V.A.R.P., H.K. and J.X. All authors have read and agreed to the published version of the manuscript.

Funding: This research received no external funding.

Institutional Review Board Statement: No ethical declarations are required as the study protocol was a part of routine diagnostic procedure, and no additional experiment was performed.

Informed Consent Statement: Not applicable.

Data Availability Statement: Genomic and meta-barcode data of the present study are submitted under bioproject number PRJNA917187.

Acknowledgments: A.C. is a Fellow of the CIFAR program Fungal Kingdom: Threats & Opportunities. A.Y. and K.J. acknowledge support by the University Grants Commission (file no. 16-9/2019) and 596/(CSIR-UGC NET JUNE 2019) Government of India (GoI), New Delhi, and Council of Scientific & Industrial Research, GoI (file no. 08/728(0001)/2019-EMR-I), respectively, for their research fellowships. J.X. acknowledges a research grant from McMaster University's Global Science Initiative (2020-03).

Conflicts of Interest: The authors declare no conflict of interest.

References

1. Satoh, K.; Makimura, K.; Hasumi, Y.; Nishiyama, Y.; Uchida, K.; Yamaguchi, H. *Candida auris* sp. nov., a novel ascomycetous yeast isolated from the external ear canal of an inpatient in a Japanese hospital. *Microbiol. Immunol.* **2009**, *53*, 41–44. [CrossRef] [PubMed]
2. Du, H.; Bing, J.; Hu, T.; Ennis, C.L.; Nobile, C.J.; Huang, G. *Candida auris*: Epidemiology, biology, antifungal resistance, and virulence. *PLoS Pathog.* **2020**, *16*, e1008921. [CrossRef] [PubMed]

3. Wang, Y.; Xu, J. Population genomic analyses reveal evidence for limited recombination in the superbug *Candida auris* in nature. *Comput. Struct. Biotechnol. J.* **2022**, *20*, 3030–3040. [CrossRef]
4. Parums, D.V. Editorial: The World Health Organization (WHO) fungal priority pathogens list in response to emerging fungal pathogens during the COVID-19 pandemic. *Med. Sci. Monit.* **2022**, *28*, e939088. [CrossRef] [PubMed]
5. Arora, P.; Singh, P.; Wang, Y.; Yadav, A.; Pawar, K.; Singh, A.; Padmavati, G.; Xu, J.; Chowdhary, A. Environmental isolation of *Candida auris* from the coastal wetlands of Andaman Islands, India. *mBio* **2021**, *12*, e03181-20. [CrossRef] [PubMed]
6. Escandón, P. Novel environmental niches for *Candida auris*: Isolation from a coastal habitat in Colombia. *J. Fungi* **2022**, *8*, 748. [CrossRef]
7. Yadav, A.; Jain, K.; Wang, Y.; Pawar, K.; Kaur, H.; Sharma, K.K.; Tripathy, V.; Singh, A.; Xu, J.; Chowdhary, A. *Candida auris* on apples: Diversity and clinical significance. *mBio* **2022**, *13*, e0051822. [CrossRef]
8. Chavez, J.; Iverson, T.; Hergert, J.; Oakeson, K.; LaCross, N.; Njoku, C.; Gorzalski, A.; Gerrity, D. *Candida auris* discovery through community wastewater surveillance during healthcare outbreak, Nevada, USA, 2022. *Emerg. Infect. Dis.* **2023**, *29*, 422–425.
9. Irinyi, L.; Roper, M.; Malik, R.; Meyer, W. Finding a needle in a haystack—In silico search for environmental traces of *Candida auris*. *Jpn. J. Infect. Dis.* **2022**, *75*, 490–495. [CrossRef]
10. Seyedmousavi, S.; Bosco, S.M.G.; de Hoog, S.; Ebel, F.; Elad, D.; Gomes, R.R.; Jacobsen, I.D.; Jensen, H.E.; Martel, A.; Mignon, B.; et al. Fungal infections in animals: A patchwork of different situations. *Med. Mycol.* **2018**, *56* (Suppl. S1), S165–S187. [CrossRef]
11. Yadav, A.; Singh, A.; Wang, Y.; Haren, M.H.V.; Singh, A.; de Groot, T.; Meis, J.F.; Xu, J.; Chowdhary, A. Colonisation and transmission dynamics of *Candida auris* among chronic respiratory diseases patients hospitalised in a chest hospital, Delhi, India: A comparative analysis of whole genome sequencing and microsatellite typing. *J. Fungi* **2021**, *7*, 81. [CrossRef] [PubMed]
12. Kathuria, S.; Singh, P.K.; Sharma, C.; Prakash, A.; Masih, A.; Kumar, A.; Meis, J.F.; Chowdhary, A. Multidrug-resistant *Candida auris* misidentified as *Candida haemulonii*: Characterization by Matrix-Assisted Laser Desorption Ionization-Time of Flight Mass Spectrometry and DNA sequencing and its antifungal susceptibility profile variability by Vitek 2, CLSI broth microdilution, and Etest Method. *J. Clin. Microbiol.* **2015**, *53*, 1823–1830. [PubMed]
13. Gweon, H.S.; Oliver, A.; Taylor, J.; Booth, T.; Gibbs, M.; Read, D.S.; Griffiths, R.I.; Schonrogge, K. PIPITS: An automated pipeline for analyses of fungal internal transcribed spacer sequences from the Illumina sequencing platform. *Methods Ecol. Evol.* **2015**, *6*, 973–980. [CrossRef] [PubMed]
14. Boratyn, G.M.; Thierry-Mieg, J.; Thierry-Mieg, D.; Busby, B.; Madden, T.L. Magic-BLAST, an accurate RNA-seq aligner for long and short reads. *BMC Bioinform.* **2019**, *20*, 405. [CrossRef]
15. Hunter, J.D. Matplotlib: A 2D Graphics Environment. *Comput. Sci. Eng.* **2007**, *9*, 90–95. [CrossRef]
16. Waskom, M.L. Seaborn: Statistical data visualization. *J. Open Source Softw.* **2021**, *6*, 3021. [CrossRef]
17. *M27-A3*; Reference Method for Broth Dilution Antifungal Susceptibility Testing of Yeasts. 3rd ed. Clinical and Laboratory Standards Institute: Wayne, PA, USA, 2008.
18. Chowdhary, A.; Prakash, A.; Sharma, C.; Kordalewska, M.; Kumar, A.; Sarma, S.; Tarai, B.; Singh, A.; Upadhyaya, G.; Upadhyay, S.; et al. A multicentre study of antifungal susceptibility patterns among 350 *Candida auris* isolates (2009–2017) in India: Role of the *ERG11* and *FKS1* genes in azole and echinocandin resistance. *J. Antimicrob. Chemother.* **2018**, *73*, 891–899. [CrossRef]
19. *M100*; Performance Standards for Antimicrobial Disk Susceptibility Testing. 32nd ed. Clinical and Laboratory Standards Institute: Wayne, PA, USA, 2022.
20. Stamatakis, A. RAxML version 8: A tool for phylogenetic analysis and post-analysis of large phylogenies. *Bioinformatics* **2014**, *30*, 1312–1313. [CrossRef]
21. Sharma, C.; Kumar, N.; Pandey, R.; Meis, J.F.; Chowdhary, A. Whole genome sequencing of emerging multidrug resistant *Candida auris* isolates in India demonstrates low genetic variation. *New Microbes New Infect.* **2016**, *13*, 77–82. [CrossRef]
22. Miller, M.A.; Pfeiffer, W.; Schwartz, T. *Creating the CIPRES Science Gateway for Inference of Large Phylogenetic Trees, in 2010 Gateway Computing Environments Workshop (GCE)*; IEEE: Piscataway, NJ, USA, 2010; pp. 1–8.
23. Letunic, I.; Bork, P. Interactive tree of life (iTOL) v4: Recent updates and new developments. *Nucleic Acids Res.* **2019**, *47*, W256–W259. [CrossRef]
24. Abastabar, M.; Haghani, I.; Ahangarkani, F.; Rezai, M.S.; Taghizadeh Armaki, M.; Roodgari, S.; Kiakojuri, K.; Al-Hatmi, A.M.S.; Meis, J.F.; Badali, H. *Candida auris* otomycosis in Iran and review of recent literature. *Mycoses* **2019**, *62*, 101–105. [CrossRef]
25. Chow, N.A.; de Groot, T.; Badali, H.; Abastabar, M.; Chiller, T.M.; Meis, J.F. Potential Fifth Clade of *Candida auris*, Iran, 2018. *Emerg. Infect. Dis.* **2019**, *25*, 1780–1781. [CrossRef] [PubMed]
26. Spruijtenburg, B.; Badali, H.; Abastabar, M.; Mirhendi, H.; Khodavaisy, S.; Sharifisooraki, J.; Taghizadeh Armaki, M.; de Groot, T.; Meis, J.F. Confirmation of fifth *Candida auris* clade by whole genome sequencing. *Emerg. Microbes Infect.* **2022**, *11*, 2405–2411. [CrossRef] [PubMed]
27. Puigdemont, A.; D'Andreano, S.; Ramió-Lluch, L.; Cuscó, A.; Francino, O.; Brazis, P. Effect of an anti-inflammatory pomegranate otic treatment on the clinical evolution and microbiota profile of dogs with otitis externa. *Vet. Dermatol.* **2021**, *32*, 158-e37. [CrossRef] [PubMed]

Journal of
Fungi

MDPI

Article

Synthesis of Altissimacoumarin D and Other Prenylated Coumarins and Their Ability to Reverse the Multidrug Resistance Phenotype in *Candida albicans*

Anna Claudia Silva [1], Daniel Clemente de Moraes [2], Denilson Costa do Carmo [1], Giselle Cristina Casaes Gomes [1], A. Ganesan [3], Rosangela Sabbatini Capella Lopes [1], Antonio Ferreira-Pereira [2,*] and Cláudio Cerqueira Lopes [1,*]

[1] Departamento de Química Analítica, Instituto de Química, Universidade Federal do Rio de Janeiro, Centro de Tecnologia, Bloco A, 508, Rio de Janeiro 21949-900, Brazil
[2] Departamento de Microbiologia Geral, Instituto de Microbiologia Paulo de Góes, Universidade Federal do Rio de Janeiro, Centro de Ciências da Saúde, Bloco I, 44, Rio de Janeiro 21941-902, Brazil
[3] School of Pharmacy, University of East Anglia, Norwich Research Park, Norwich NR4 7TJ, UK
* Correspondence: apereira@micro.ufrj.br (A.F.-P.); claudioc@iq.ufrj.br (C.C.L.)

Abstract: Azoles are the main antifungal agents employed in clinical practice to treat invasive candidiasis. Nonetheless, their efficacy is limited by fungal resistance mechanisms, mainly the overexpression of efflux pumps. Consequently, candidiasis has a worrisome death rate of 75%. One potential strategy to overcome efflux-mediated resistance is to inhibit this process. *Ailanthus altissima* is a Chinese tree that produces several active substances, including altissimacoumarin D. Due to the low yield of its extraction and the need to search for new drugs to treat candidiasis, this study aimed to synthesize altissimacoumarin D and its analogues, as well as evaluating their ability to reverse the resistance phenotype of *Candida albicans*. Coumarin isofraxidin was prepared via total synthesis through a solvent-free Knoevenagel condensation as the key step. Isofraxidin and other commercially available coumarins were alkylated with prenyl or geranyl groups to yield the natural product altissimacoumarin D and seven analogues. The antifungal activity of the coumarins and their ability to reverse the fungal resistance phenotype were assessed using microbroth methodologies. Toxicity was evaluated using erythrocytes and an in silico prediction. All compounds improved the antifungal activity of fluconazole by inhibiting efflux pumps, and ACS47 and ACS50 were the most active. None of the coumarins were toxic to erythrocytes. In silico predictions indicate that ACS47 and ACS50 may be safe for human use. ACS47 and ACS50 are promising candidates when used as adjuvants in the antifungal therapy against *C. albicans*-resistant strains.

Keywords: altissimacoumarin D; *Candida* spp.; coumarins; efflux transporters; isofraxidin; multidrug resistance

1. Introduction

Fungi belonging to the *Candida* genus can be found within the human microbiome, at sites such as skin and urinary tract [1]. Nonetheless, these microorganisms may invade inner organs and cause life-threatening infections, mainly in immunocompromised patients. In this context, *Candida* spp. is the most significant fungal pathogen in humans, and its infection has mortality rates of 46–75% [2].

Currently, treating invasive fungal infections relies on a limited number of antifungal agents—azoles, polyenes and echinocandins—and the increasing incidence of resistance to them jeopardizes the success of therapy [3]. Azole drugs, such as fluconazole, itraconazole and ketoconazole, are first-line treatments and are subject to several mechanisms of resistance by *Candida* spp. [4].

The most important treatment involves the overexpression of efflux pumps within the plasma membrane [5]. These transporters may belong to the ATP-Binding Cassette

(ABC) superfamily, which uses ATP hydrolysis as energy source to extrude drugs from the cytoplasm to the extracellular milieu [6], or to the Major Facilitator Superfamily (MFS), which uses proton gradient to transport the drugs [7] and confers a multidrug resistance (MDR) phenotype. Overcoming efflux-pump-mediated resistance is thereby an important strategy for combating fungal resistance to current drugs [8].

Ailanthus altissima is a Chinese native tree with a history of use in traditional herbal medicines for a variety of ailments that range from vaginal infections to epilepsy. Several compounds can be extracted and isolated from the plant, including approximately one hundred coumarins. These substances mostly show antitumor and deacetylating activity. Nonetheless, they are not the most studied group of compounds found in *A. altissima* and require further investigation [9]. Coumarins, and especially prenylated coumarins, have already been studied and have antibacterial [10,11], antifungal [12], antileishmanial [13], anticancer [14], anti-inflammatory and antiviral [15] activities.

In 2012, a group of coumarins was isolated for the first time from *A. altissima*, and their deacetylating activity was demonstrated. However, no other studies have been performed since then, which was probably a consequence of the difficulty of obtaining such compounds, considering that 3 kg of the tree bark provides 6.5 to 22 mg of the coumarins [16]. In a previous study by our group, we were able to perform the first total synthesis of altissimacoumarin D (Figure 1), the most active compound isolated from *A. altissima* by Dao et al. [17].

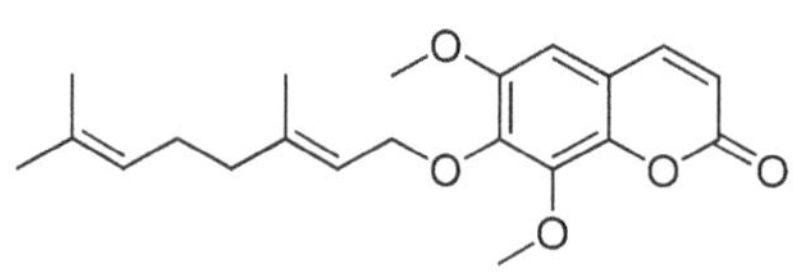

Figure 1. Structure of altissimacoumarin D (ACS51).

In this study, we aimed to improve the total synthesis of altissimacoumarin D by decreasing the number of steps and toxic reagents used, avoiding high temperatures (e.g., ethylene glycol reflux) and the use of anhydrous solvents. Analogues were also synthesized, and we demonstrated that these compounds have the ability to reverse the MDR phenotype in *Candida albicans*.

2. Experimental Section

2.1. General Experiment Procedure

All reagents were obtained commercially (Sigma-Aldrich Co., St. Louis, MO, USA) and used without further purification, except fluconazole which was obtained commercially from the university pharmacy (UFJF, Juiz de Fora—MG, Brazil). Analytical thin-layer chromatography was carried out on 0.25 mm silica gel 60 F254 plates, and compounds were visualized using UV irradiation or staining with 2,4-dinitrophenylhydrazine. Nuclear magnetic resonance (NMR) spectra were obtained on a Bruker Avance-III spectrometer (Bruker Corporation, Billerica, MA, USA) (Figures S1–S16). The splitting patterns are reported as s (singlet), d (doublet), t (triplet), q (quartet) and m (multiplet). Fluconazole stock solutions were prepared in distilled water, sterilized by filtration (0.22 μm), and maintained at −20 °C.

2.2. Synthetic Procedures

2.2.1. 3,5-Dibromo-2,4-dihydroxybenzaldehyde

This compound was prepared as previously described [17]. ^{1}H NMR (400 MHz, DMSO-d_6) δ 7.96 (s, 1H), 9.78 (s, 1H); ^{13}C NMR (100 MHz, DMSO-d_6) δ 100.1, 101.9, 116.3, 135.7, 158.1, 158.7, 193.5.

2.2.2. 2,4-Dihydroxy-3,5-dimethoxybenzaldehyde

This compound was prepared as previously described [17]. ^{1}H NMR (400 MHz, $CDCl_3$) δ 3.98 (s, 6H), 7.28 (s, 1H), 8.0 (s, 1H); ^{13}C NMR (100 MHz, $CDCl_3$) δ 56.6, 60.9, 109.0, 112.9, 134.6, 141.3, 147.1, 151.7, 194.6.

2.2.3. 7-Hydroxy-6,8-dimethoxy-2H-chromen-2-one (isofraxidin)

2,4-dihydroxy-3,5-dimethoxybenzaldehyde (1 eq), Meldrum's acid (1.2 eq) and ammonium carbonate (0.2 eq) were placed in a sealed reaction tube and stirred at 75 °C for 2 h. The reaction was monitored using thin-layer chromatography with a solution of iron (III) chloride to determine the consumption of the starting aldehyde. Then, the temperature was increased to 140 °C for decarboxylation. This step was also monitored using thin-layer chromatography. After completion, a small amount of acetone was added to solubilize the reaction mixture, which was then transferred into a saturated solution of sodium bicarbonate. The product was extracted using ethyl acetate (5 × 20 mL). The organic layer was washed with brine, dried with Na_2SO_4, filtered, and the solvent evaporated was in vacuo [18]. ^{1}H NMR (400 MHz, $CDCl_3$) δ 3.86 (s, 3H), 4.01 (s, 3H), 6.21 (d, *J* 9.0 Hz, 1H), 6.59 (s, 1H), 7.54 (d, *J* 9.0 Hz, 1H); ^{13}C NMR (100 MHz, DMSO-d_6) δ 56.2, 60.8, 105.9, 109.5, 112.8, 134.2, 144.5, 145.9, 146.9, 149.6, 164.2. Yield: 60%.

2.2.4. €-7-((3,7-Dimethylocta-2,6-dien-1-yl)oxy)-6,8-dimethoxy-2H-chromen-2-one (ACS51)

K_2CO_3 (1.2 eq) and geranyl bromide (1.0 eq) were added to a solution of isofraxidin (1.0 eq) in acetone (5 mL), and the mixture was stirred under reflux until completion. The product was extracted using ethyl acetate (5 × 20 mL). The organic layer was washed with brine, dried with Na_2SO_4, filtered, and the solvent was evaporated in vacuo. ^{1}H NMR (400 MHz, $CDCl_3$) δ 1.59 (s, 3H), 1.66 (s, 3H), 1.69 (s, 3H), 2.05 (m, 4H), 3.89 (s, 3H), 4.03 (s, 3H), 4.68 (d, *J* 7 Hz, 2H), 5.06 (t, *J* 8 Hz, 1H), 5.55 (t, *J* 8 Hz, 1H), 6.35 (d, *J* 8 Hz, 1H), 6.66 (s, 1H), 7.60 (d, *J* 12 Hz, 1H); ^{13}C NMR (100 MHz, $CDCl_3$) δ 16.4, 17.7, 25.7, 26.4, 39.6, 56.3, 61.7, 70.3, 103.5, 114.4, 115.1, 119.6, 123.8, 131.7, 141.8, 142.5, 143.0, 143.5, 144.9, 150.7, 160.6. Yield: 70%.

2.2.5. 6,8-Dimethoxy-7-((3-methylbut-2-en-1-yl)oxy)-2H-chromen-2-one (ACS50)

K_2CO_3 (1.2 eq) and prenyl bromide (1.0 eq) were added to a solution of isofraxidin (1.0 eq) in acetone (5 mL) and the mixture was stirred under reflux until completion. The product was extracted using ethyl acetate (5 × 20 mL). The organic layer was washed with brine, dried with Na_2SO_4, filtered, and the solvent evaporated in vacuo. ^{1}H NMR (400 MHz, $CDCl_3$) δ 1.71 (s, 3H), 1.77 (s, 3H), 3,89 (s, 3H), 4.03 (s, 3H), 4.65 (d, *J* 8 Hz, 2H), 5.56 (t, *J* 8 Hz, 1H), 6.35 (d, *J* 8 Hz, 1H), 6.67 (s, 1H), 7.63 (d, *J* 8 Hz, 1H); ^{13}C NMR (100 MHz, $CDCl_3$) δ 18.2, 26.0, 56.5, 62.0, 70.5, 103.8, 114.6, 115.3, 120.1, 139.5, 142.1, 143.2, 143.7, 150.9, 160.8. Yield: 77%.

2.2.6. (E)-7-((3,7-Dimethylocta-2,6-dien-1-yl)oxy)-2H-chromen-2-one (ACS48)

K_2CO_3 (1.2 eq) and geranyl bromide (1.0 eq) were added to a solution of umbelliferone (1.0 eq) in acetone (5 mL), and the mixture was stirred under reflux until completion. The product was extracted using ethyl acetate (5 × 20 mL). The organic layer was washed with brine, dried with Na_2SO_4, filtered, and the solvent was evaporated in vacuo. ^{1}H NMR (400 MHz, $CDCl_3$) δ 1.62 (s, 3H), 1.68 (s, 3H), 1.77 (s, 3H), 2.11 (m, 4H), 4.61 (d, J 6 Hz, 2H), 5.01 (m, 1H), 5.48 (m, 1H), 6.26 (d, *J* 8 Hz, 1H), 6.85 (m, 2H), 7.36 (d, *J* 8 Hz, 1H), 7.65 (d, *J* 9 Hz, 1H). ^{13}C NMR (100 MHz, $CDCl_3$) δ 17.0, 17.9, 25.9, 26.5, 39.7, 65.7, 101.8, 112.6, 113.2, 113.5, 118.6, 123.8, 128.9, 132.2, 142.6, 143.7, 156.1, 161.5, 162.4. Yield: 82%.

2.2.7. 7-((3-Methylbut-2-en-1-yl)oxy)-2H-chromen-2-one (ACS47)

K_2CO_3 (1.2 eq) and prenyl bromide (1.0 eq) were added to a solution of umbelliferone (1.0 eq) in acetone (5 mL), and the mixture was stirred under reflux until completion. The product was extracted using ethyl acetate (5 × 20 mL). The organic layer was washed with brine, dried with Na_2SO_4, filtered, and the solvent was evaporated in vacuo. ^{1}H NMR (400 MHz, $CDCl_3$) δ 1.77 (s, 3H), 1.82 (s, 3H), 4.59 (d, *J* 7 Hz, 2H), 5.48 (t, *J* 6 Hz, 1H), 6.25 (d, *J* 9 Hz, 1H), 6.86 (m, 2H), 7.36 (d, *J* 8 Hz, 1H). ^{13}C NMR (100 MHz, $CDCl_3$) δ 18.3, 25.8, 65.4, 101.6, 112.4, 113.0, 113.2, 118.6, 128.7, 139.3, 143.5, 155.9, 161.3, 162.1. Yield: 79%.

2.2.8. (E)-7-((3,7-Dimethylocta-2,6-dien-1-yl)oxy)-4-methyl-2H-chromen-2-one (ACS52)

K_2CO_3 (1.2 eq) and geranyl bromide (1.0 eq) were added to a solution of 4-methylumbelliferone (1.0 eq) in acetone (5 mL), and the mixture was stirred under reflux until completion. The product was extracted using ethyl acetate (5 × 20 mL). The organic layer was washed with brine, dried with Na_2SO_4, filtered, and the solvent was evaporated in vacuo. ^{1}H NMR (400 MHz, $CDCl_3$) δ 1.62 (s, 3H), 1.68 (s, 3H), 1.77 (s, 3H), 2.11 (m, 4H), 2.41 (s, 3H), 4.61 (d, *J* 6 Hz, 2H), 5.01 (m, 1H), 5.48 (m, 1H), 6.14 (s, 1H), 6.83 (s, 1H), 6.87 (d, *J* 9 Hz,1H), 7.50 (d, *J* 9 Hz, 1H). ^{13}C NMR (100 MHz, $CDCl_3$) δ 16.8, 17.7, 18.7, 25.7, 26.3, 39.5, 65.5, 101.6, 111.9, 112.9, 113.5, 118.5, 123.6, 125.4, 132.0, 142.3, 152.6, 155.3, 161.4, 162.0. Yield: 91%.

2.2.9. 4-Methyl-7-((3-methylbut-2-en-1-yl)oxy)-2H-chromen-2-one (ACS54)

K_2CO_3 (1.2 eq) and prenyl bromide (1.0 eq) were added to a solution of 4-methylumbelliferone (1.0 eq) in acetone (5 mL), and the mixture was stirred under reflux until completion. The product was extracted using ethyl acetate (5 × 20 mL). The organic layer was washed with brine, dried with Na_2SO_4, filtered, and the solvent was evaporated in vacuo. ^{1}H NMR (400 MHz, $CDCl_3$) δ 1.78 (s, 3H), 1.82 (s, 3H), 2.40 (s, 3H), 4.59 (d, *J* 7 Hz, 2H), 5.48 (t, *J* 6 Hz, 1H), 6.13 (s, 1H), 6.83 (s, 1H), 6.86 (d, *J* 7 Hz, 1H), 7.48 (d, *J* 9 Hz, 1H). ^{13}C NMR (100 MHz, $CDCl_3$) δ 18.3, 18.7, 25.8, 65.4, 101.6, 111.8, 112.9, 113.5, 118.7, 125.5, 139.2, 152.6, 155.2, 161.4, 161.9. Yield: 86%.

2.2.10. (E)-7-((3,7-Dimethylocta-2,6-dien-1-yl)oxy)-3,4,8-trimethyl-2H-chromen-2-one (ACS55)

K_2CO_3 (1.2 eq) and prenyl bromide (1.0 eq) were added to a solution of 7-hydroxy-3,4,8-trimethylcoumarin (1.0 eq) in acetone (5 mL), and the mixture was stirred under reflux until completion. The product was extracted using ethyl acetate (5 × 20 mL). The organic layer was washed with brine, dried with Na_2SO_4, filtered, and the solvent was evaporated in vacuo. ^{1}H NMR (400 MHz, $CDCl_3$) δ 1.59 (s, 3H), 1.65 (s, 3H), 1.73 (s, 3H), 2.08 (m, 4H), 2.16 (s, 3H), 2.29 (s, 3H), 2.34 (s, 3H), 4.60 (d, *J* 6 Hz, 2H), 5.07 (m, 1H), 5.46 (m, 1H), 6.79 (d, *J* 9 Hz,1H), 7.36 (d, *J* 9 Hz, 1H). ^{13}C NMR (100 MHz, $CDCl_3$) δ 8.5, 13.4, 15.3, 17.0, 17.9, 25.9, 26.5, 39.7, 65.9, 108.2, 114.3, 114.5, 118.8, 119.7, 122.1, 123.9, 132.1, 141.4, 146.6, 151.4, 158.8, 162.9. Yield: 85%.

2.2.11. 3,4,8-Trimethyl-7-((3-methylbut-2-en-1-yl)oxy)-2H-chromen-2-one (ACS56)

K_2CO_3 (1.2 eq) and geranyl bromide (1.0 eq) were added to a solution of 7-hydroxy-3,4,8-trimethylcoumarin (1.0 eq) in acetone (5 mL), and the mixture was stirred under reflux until completion. The product was extracted using ethyl acetate (5 × 20 mL). The organic layer was washed with brine, dried with Na_2SO_4, filtered, and the solvent was evaporated in vacuo. ^{1}H NMR (400 MHz, $CDCl_3$) δ 1.71 (s, 3H), 1.77 (s, 3H), 2.18 (s, 3H), 2,31 (s, 3H), 2.36 (s, 3H), 4.60 (d, *J* 7 Hz, 2H), 5.50 (t, *J* 6 Hz, 1H), 6.83 (d, *J* 9 Hz, 1H), 7.40 (d, *J* 9 Hz, 1H). ^{13}C NMR (100 MHz, $CDCl_3$) δ 8.3, 13.2, 15.1, 18.3, 25.8, 65.6, 108.0, 114.1, 114.3, 118.6, 119.6, 121.9, 138.0, 146.4, 151.2, 158.6, 162.7. Yield: 83%.

2.3. *Biological Assays*

2.3.1. Strains

In this study, four mutant strains of *S. cerevisiae* were used. These strains were kindly provided by Dr. Richard Cannon and Dr. Brian Monk (University of Otago, Dunedin, New Zealand). Firstly, a null mutant was constructed by deleting genes related to ABC transporters (Pdr5p, Yor1p, Snq2p, Pdr10p, Pdr11p, and Ycf1p). Then, three fluconazole-resistant strains were created by heterologous expression of transporters originally found in *Candida* spp., as summarized in Table 1 [19]. Also, two fluconazole-resistant *C. albicans* clinical isolates were used. While 95-142 overexpresses the ABC transporters, CaCdr1p and CaCdr2p [5], the PRI strain overexpresses the MFS transporter CaMdr1p [20]. The 95-142 strain was kindly provided by Dr. Theodore White (University of Missouri, Kansas City, MO, USA). Additionally, one fluconazole-sensitive *C. albicans* clinical isolate was used (ATCC 10231; American Type Culture Collection™, Manassas, VA, USA). The strains were grown in YPD medium (2% glucose, 1% yeast extract, 2% peptone) at 30 °C (*S. cerevisiae*) or 37 °C (*C. albicans*) via agitation and were harvested in the exponential phase of growth whenever experiments were about to be performed.

Table 1. Strains used in this study.

Strain	Efflux Transporter	Family	Species
Null mutant	None	-	*S. cerevisiae*
CaCdr1p+	CaCdr1p	ABC	*S. cerevisiae*
CaCdr2p+	CaCdr2p	ABC	*S. cerevisiae*
CaMdr1p+	CaMdr1p	MFS	*S. cerevisiae*
95-142	CaCdr1p and CaCdr2p	ABC	*C. albicans*
PRI	CaMdr1p	MFS	*C. albicans*
ATCC 10231	None	-	*C. albicans*

2.3.2. Antifungal Activity of Coumarins against *C. albicans* and *S. cerevisiae* Strains

Briefly, 5×10^3 cells/mL of *C. albicans* strains (95-142, PRI and ATCC 10231) and 2×10^4 cells/mL of the *S. cerevisiae* null mutant were incubated in RPMI-1640 (Sigma-Aldrich Co., St. Louis, MO, USA) at 37 °C or YPD medium at 30 °C, respectively, for 48 h via agitation (75 rpm) in the presence of 100 µg/mL of the compounds [21]. Cell growth was measured at 600 nm (Fluostar Optima, BMG Labtech, Offenburg, Germany).

2.3.3. Antifungal Activity of Coumarins Combined with Fluconazole against *S. cerevisiae* Strains

Briefly, 2×10^4 cells/mL of *S. cerevisiae* mutant strains (CaCdr1p+, CaCdr2p+, and CaMdr1p+) were incubated in YPD medium at 30 °C for 48 h using agitation (75 rpm), in the presence of 100 µg/mL of the coumarins with or without fluconazole at sub-inhibitory concentrations (125 µg/mL, 15.6 µg/mL and 15.6 µg/mL, respectively) [21]. Cell growth was measured at 600 nm (Fluostar Optima, BMG Labtech, Offenburg, Germany).

2.3.4. Synergism Evaluation through Checkerboard Assay

This assay was conducted with the compounds active against *S. cerevisiae* in the presence of fluconazole. Briefly, 2×10^4 cells/mL of *S. cerevisiae* mutant strains (CaCdr1p+, CaCdr2p+, and CaMdr1p+) were incubated in YPD medium at 30 °C for 48 h with agitation (75 rpm), in the presence of serial concentrations of the coumarins (100–6.25 µg/mL) and fluconazole (250–3.91 µg/mL). Cell growth was measured at 600 nm (Fluostar Optima, BMG Labtech, Offenburg, Germany). Serial dilutions of the coumarins and fluconazole alone were made to obtain the minimum inhibitory concentration (MIC) of each compound. The interaction between the coumarins and fluconazole was assessed using the fractional inhibitory concentration index (FICI). The FICI is calculated by summing the fractional inhibitory concentration (FIC) of each drug, and the FIC is the ratio of the MIC of a drug combined with a second drug and the MIC of a drug alone. Synergistic, additive, indifferent, and antagonistic interactions correspond to FICI ≤ 0.5, >0.5–1.0, 1.0–4.0, and >4.0, respectively [21].

After evaluating the association between coumarins and fluconazole against *S. cerevisiae* strains, the combinations that presented either additive or synergistic interactions were tested against the growth of *C. albicans*. The compounds active against CaCdr1p+ or CaCdr2p+ were tested against the 95-142 strain, while the compounds active against CaMdr1p+ were tested against the PRI strain. Briefly, 5×10^3 cells/mL of *C. albicans* resistant strains was incubated in RPMI-1640 medium at 37 °C for 48 h with agitation

(75 rpm) in the presence of serial concentrations of the coumarins (100–6.25 μg/mL) and fluconazole (1000–15.6 μg/mL). Cell growth was measured at 600 nm (Fluostar Optima, BMG Labtech, Offenburg, Germany), and the FICI values were calculated using the aforementioned method.

2.3.5. Erythrocyte Viability

The effect of the compounds on erythrocyte viability was assessed as described elsewhere [22]. Cells were washed three times and resuspended in phosphate-buffered saline (PBS) to a final of concentration of 2% v/v. Then, cells were incubated in the presence of the compounds at 100 μg/mL for 60 min at 37 °C. Afterward, cells were harvested by centrifugation at 3000× *g* for 5 min and 1 mL of supernatant was transferred to glass cuvettes. The absorbance of hemoglobin released from erythrocytes was measured at 540 nm (SP220, Biospectro, Curitiba, Brazil). Controls of 100% and 0% hemolysis were performed by incubating the cells in PBS in the presence or absence of 1% Triton X-100, respectively. Also, controls with DMSO and acetonitrile were performed.

2.3.6. In Silico Toxicity Prediction

The toxicity of the coumarins was predicted in silico using OSIRIS Property Explorer [23], which provides information regarding mutagenic, tumorigenic, irritant, and reproductive effects, and GUSAR (General Unrestricted Structure–Activity Relationships), which provides a LD_{50} (lethal dose to 50% of a given population) of the substances in rats, considering four routes of administration (intraperitoneal, intravenous, oral, and subcutaneous) [24].

2.3.7. Statistical Analysis

All the experiments were performed at least three times, and the results were expressed as mean ± standard deviation. Data were analyzed using Student's t test, and p values lower than 0.05 were considered significant.

3. Results

3.1. Synthesis

The synthesis of altissimacoumarin D is summarized in Figure 2. This coumarin was first isolated by Dao et al. in 2012 [16], and the first total synthesis performed by our group [17] involved the alkylation of isofraxidin, another natural occurring coumarin. In this study, our goal was to improve that first synthesis via the most crucial step, the formation of the coumarin ring. The first two steps of the synthesis, which lead to the formation of the methoxylated intermediate, were kept the same, since the yields were already improved, and the purification process was simple.

Attempts were made to form the coumarin ring in a way that would be more efficient, greener and easier to purify. Then, the reaction was performed as described by Schijndel et al. [18], with slight modifications. Instead of forming the coumarin ring using malonic acid, 2-hydroxybenzadehyde and ammonium bicarbonate, we used ammonium carbonate, 2,4-dihydroxy-3,5-dimethoxybenzaldehyde and Meldrum's acid with no solvent in this study. Isofraxidin was obtained with a yield of 60%.

After obtaining isofraxidin, the next step was to obtain altissimacoumarin D via an alkylation. In our previous study, this step was performed via a Mitsunobu reaction that, although efficient, required the use of a dry solvent, a 24 h reaction, and made the purification of the product challenging. Thus, in this study, we performed the alkylation via a simpler methodology using acetone, potassium carbonate and geranyl bromide; altissimacoumarin D was obtained with a yield of 70%.

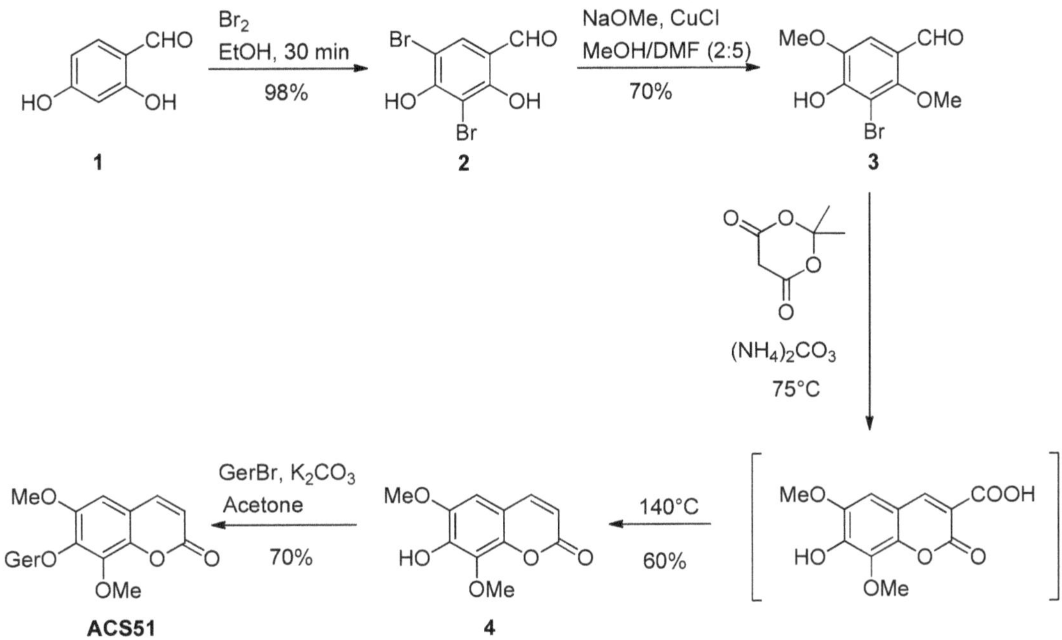

Figure 2. Total synthesis of altissimacoumarin D in four steps.

Since only one biological test was performed with altissimacoumarin D [16], and the compound was shown to be promising due to its prenylated coumarin nature, seven other coumarins with different substitution patterns were synthesized, varying the alkyl chain between geranyl and prenyl. Figure 3 shows the structures of the analogues of altissimacoumarin D that were synthesized in this study.

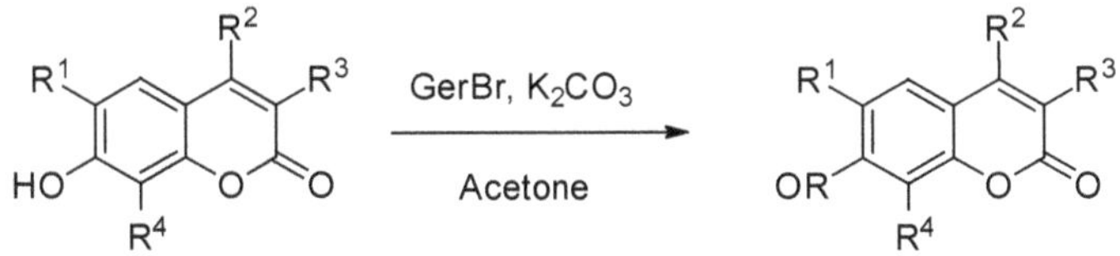

Compound	OR	R¹	R²	R³	R⁴
ACS47	OPren	H	H	H	H
ACS48	OGer	H	H	H	H
ACS50	OPren	OMe	H	H	OMe
ACS52	OGer	H	Me	H	H
ACS54	OPren	H	Me	H	H
ACS55	OPren	H	Me	Me	Me
ACS56	OGer	H	Me	Me	Me

Figure 3. Analogues of altissimacoumarin D obtained through the alkylation of commercially available compounds. OMe—methoxy group; Me—methyl group; OGer—geranyl group; OPren—prenyl group.

3.2. Antifungal Activity of Coumarins against C. albicans and S. cerevisiae Strains

Altissimacoumarin D and the coumarin analogues were screened for antifungal activity against *C. albicans* and *S. cerevisiae* strains (Figure 4).

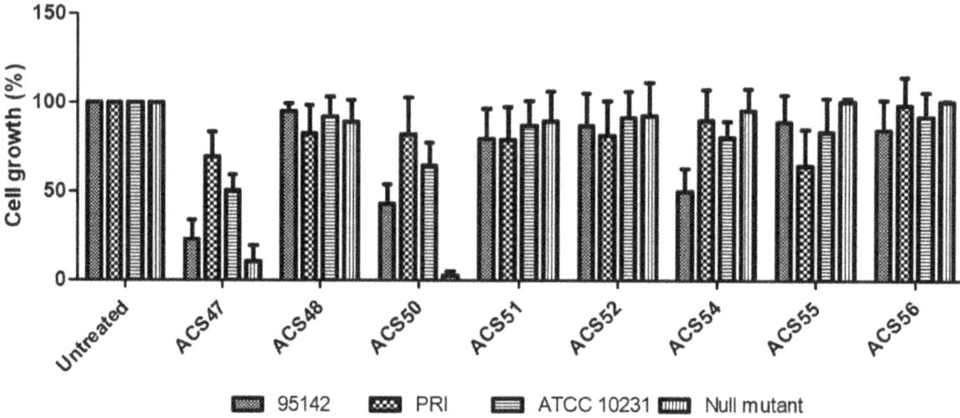

Figure 4. Antifungal activity screening against *C. albicans* and *S. cerevisiae* strains. Cells (95-142, PRI, ATCC 10231 and Null mutant) were incubated with 100 μg/mL of coumarins for 48 h. Cell growth was measured at 600 nm. Data are expressed as mean ± standard deviation of three independent experiments.

While the natural product was inactive at the tested concentration, ACS47, ACS50, and ACS54 at 100 μg/mL presented antifungal activity. ACS47 inhibited the growth of 95-142, PRI, ATCC 10231, and Null mutant strains by 77.2%, 30.4%, 49.5%, and 89.8%, respectively. ACS50 inhibited the growth of 95-142, ATCC 10231 and Null mutant strains by 57%, 35.5%, and 97.6%, respectively. ACS54 inhibited the growth of the 95-142 strain by 50.2%.

3.3. Antifungal Activity of Coumarins Combined with Fluconazole against S. cerevisiae Strains

To assess if the compounds inhibit *C. albicans* efflux pumps, their antifungal activity alone and combined with fluconazole were evaluated against *S. cerevisiae* mutant strains that overexpress these transporters. Compounds ACS48, ACS51, ACS52, ACS54, ACS55 and ACS56 did not exhibit antifungal activity alone. On the other hand, ACS47 inhibited the growth of CaCdr1p+ and CaCdr2p+ by 49.6% and 50.3%, respectively, and ACS50 inhibited the growth of CaCdr2p+ by 26.5%. However, in combination with fluconazole, all compounds, except for ACS55, inhibited the growth of CaMdr1p+ by 93.6–99.1%. Moreover, the combination of ACS47 with fluconazole inhibited the growth of CaCdr1p+ and CaCdr2p+ by 82.5% and 94.7%, respectively, while the combination of ACS50 and fluconazole inhibited the growth of CaCdr2p+ by 83.2% (Figure 5).

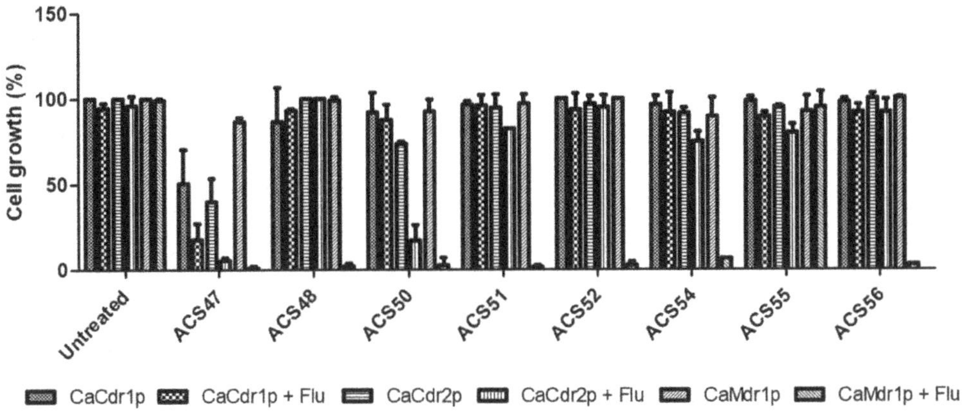

Figure 5. Antifungal activity of coumarins combined with fluconazole against *S. cerevisiae* strains. Cells (CaCdr1p+, CaCdr2p+, and CaMdr1p+) were incubated with 100 µg/mL of coumarins and sub-inhibitory concentrations of fluconazole for 48 h. Cell growth was measured at 600 nm. Data are expressed as the mean ± standard deviation of three independent experiments.

3.4. Synergism Evaluation through Checkerboard Assay

Since the coumarins presented antifungal activity when combined with fluconazole, a checkerboard assay was performed to determine the type of interaction between the compounds. Firstly, the *S. cerevisiae* mutant strains were used. As shown in Table 2, ACS47 at 50 µg/mL was synergistic with fluconazole against CaCdr1p+ and CaCdr2p+, reducing the fluconazole MIC by 4–8 fold. The interaction between ACS50 (at 25 µg/mL) and fluconazole was classified as additive, and the coumarin halved fluconazole MIC against CaCdr2p+. All the coumarins were able to diminish fluconazole MIC against CaMdr1p+. Compounds ACS47–54 displayed synergistic activity with fluconazole, with FICI values ranging from 0.188 to 0.500. ACS55 and ACS56 presented an additive interaction with fluconazole.

Table 2. Checkerboard assays of coumarins against *S. cerevisiae* mutant strains.

Strain and Compound	Compound (µg/mL)			Fluconazole (µg/mL)				
	MIC^a	MIC^b	FIC	MIC^a	MIC^b	FIC	FICI	Outcome
CaCdr1p+								
ACS47	>100	50	0.250	>250	62.5	0.125	0.375	S
CaCdr2p+								
ACS47	>100	50	0.250	62.5	15.6	0.250	0.500	S
ACS50	>100	25	0.125	62.5	31.3	0.500	0.625	A
CaMdr1p+								
ACS47	>100	25	0.125	62.5	3.91	0.063	0.188	S
ACS48	>100	50	0.250	62.5	7.81	0.125	0.375	S
ACS50	>100	25	0.125	62.5	15.6	0.250	0.375	S
ACS51	>100	50	0.250	62.5	15.6	0.250	0.500	S
ACS52	>100	12.5	0.063	62.5	15.6	0.250	0.313	S
ACS54	>100	12.5	0.063	62.5	7.81	0.125	0.188	S
ACS55	>100	100	0.500	62.5	31.3	0.500	1.000	A
ACS56	>100	100	0.500	62.5	15.6	0.250	0.750	A

MIC^a, MIC of compound alone; MIC^b, MIC of combined compound; FIC, fractional inhibitory concentration; FICI, fractional inhibitory concentration index; S, synergism; A, additive.

Due to these positive results, the checkerboard assay was also performed using *C. albicans* strains. All the compounds were tested against PRI because this strain overexpresses CaMdr1p. Furthermore, ACS47 and ACS50 were tested against 95-142, a *C. albicans* strain that overexpresses both CaCdr1p and CaCdr2p. At 50 µg/mL, ACS 47 and ACS50 reduced fluconazole MIC against 95-142 by 4–8 fold, with FICI values of 0.375 and 0.500, respectively. Against the PRI strain, only ACS47 and ACS50 showed a combined activity with fluconazole. Both compounds reduced the fluconazole MIC from 1000 µg/mL to 15.6 µg/mL (Table 3).

Table 3. Checkerboard assays of coumarins against fluconazole-resistant clinical strains of *Candida albicans*.

Strain and Compound	Compound (µg/mL)			Fluconazole (µg/mL)				
	MIC[a]	MIC[b]	FIC	MIC[a]	MIC[b]	FIC	FICI	Outcome
95-142								
ACS47	>100	50	0.250	500	62.5	0.125	0.375	S
ACS50	>100	50	0.250	500	125	0.250	0.500	S
PRI								
ACS47	>100	100	0.500	1000	15.6	0.016	0.516	A
ACS48	>100	>100	1	1000	1000	1	2	I
ACS50	>100	100	0.500	1000	15.6	0.016	0.516	A
ACS51	>100	>100	1	1000	1000	1	2	I
ACS52	>100	>100	1	1000	1000	1	2	I
ACS54	>100	>100	1	1000	1000	1	2	I
ACS55	>100	>100	1	1000	1000	1	2	I
ACS56	>100	>100	1	1000	1000	1	2	I

MIC[a], MIC of compound alone; MIC[b], MIC of combined compound; FIC, fractional inhibitory concentration; FICI, fractional inhibitory concentration index; I, indifferent; A, additive.

3.5. Erythrocyte Viability

At 100 µg/mL, the compounds exerted 0.8–6.6% of hemolysis. The lowest hemolytic compounds were ACS47 and ACS50, which presented results comparable to PBS and DMSO controls (Figure 6).

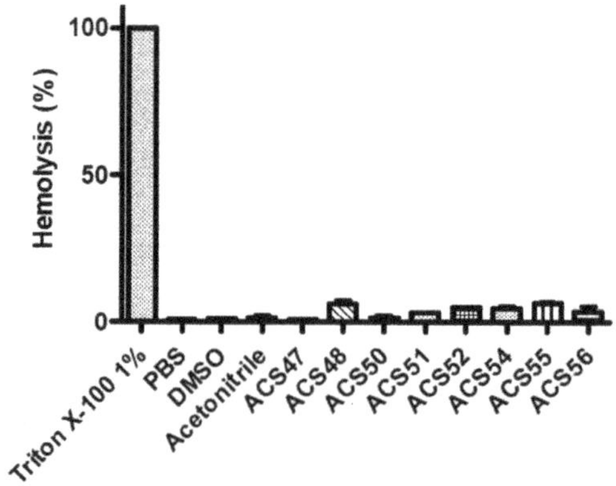

Figure 6. Hemolysis assay. Erythrocytes were incubated in the presence of coumarins, and the absorbance of released hemoglobin was measured at 540 nm. Triton X-100 and PBS were used as controls of 100% and 0% hemolysis, respectively. DMSO and acetonitrile were also used as controls. Data are expressed as mean ± standard deviation of three independent experiments. PBS: phosphate-buffered saline. DMSO: dimethyl sulfoxide.

3.6. *In Silico Toxicity Prediction*

The toxicity of the prenylated coumarins was predicted in silico using two free software programs. According to OSIRIS Property Explorer, ACS47 and ACS50 were predicted not to be mutagenic, tumorigenic and irritant; they do not have any reproductive effects on humans, a toxicity profile similar to fluconazole. ACS48 and ACS51 may be irritants, while ACS54 may induce reproductive effects. Also, ACS52, ACS55, and ACS56 may be irritant and exert reproductive effects. None of the compounds are predicted to be mutagenic or tumorigenic (Table 4).

Table 4. Toxicity risk prediction by Osiris Property Explorer.

Compound	Mutagenic	Tumorigenic	Irritant	Reproductive Effect
ACS47	No	No	No	No
ACS48	No	No	Yes	No
ACS50	No	No	No	No
ACS51	No	No	Yes	No
ACS52	No	No	Yes	Yes
ACS54	No	No	No	Yes
ACS55	No	No	Yes	Yes
ACS56	No	No	Yes	Yes

Data obtained by the GUSAR software is summarized in Table 5. Coumarins intraperitoneal, intravenous, oral, and subcutaneous LD_{50} ranged from 333.1 to 581 mg/kg, 53.41 to 121.6 mg/kg, 1057 to 2690 mg/kg, and 845.7 to 2462 mg/kg, respectively. Fluconazole LD_{50} ranged from 200.4 mg/kg to 708 mg/kg.

Table 5. Rat acute toxicity prediction by GUSAR (mg/kg).

Compound	Rat IP LD_{50}	Rat IV LD_{50}	Rat Oral LD_{50}	Rat SC LD_{50}
ACS47	476.4	65.99	1551	2462
ACS48	456.5	90.71	2157	1234
ACS50	426.8	93.38	1057	1461
ACS51	525	121.6	1401	845.7
ACS52	401.5	83.72	1979	918.6
ACS54	581.3	66.26	1470	2452
ACS55	333.1	53.41	2445	1116
ACS56	351.9	74.89	2690	948.3
Fluconazole	708	200.4	584.4	511.3

4. Discussion

4.1. *Synthesis of Prenylated Coumarins*

Natural products have been extensively explored as potential sources of new drugs, due to their wide variety of pharmacological activities. The Chinese tree *A. altissima*, for example, possesses compounds with anti-inflammatory [25], anticancer [26], and neuroprotective [27] properties. Also, Li et al. (2021) isolated an alkaloid with antifungal activity against *Fusarium oxysporum* [28], and Dao et al. (2012) purified terpenylated coumarins with potential anticancer activity from *A. altissima* [16]. Coumarins are considered privileged scaffolds for the synthesis of potential drugs with higher yields when compared to isolation from natural sources [29]. The presence of terpenylated coumarins in *A. altissima*, the feasibility of their synthesis, their potential pharmacological activity, and the need to discover substances that reverse antifungal resistance in *Candida* spp. encouraged the synthesis and evaluation of antifungal activity of altissimacoumarin D, the main compound of the class found in *A. altissima* and its analogues. In the present study, isofraxidin was efficiently prepared in a one-pot solvent-free reaction. Also, the Mitsunobu reaction, which was a different approach from the previous synthesis of altissimacoumarin D that we reported, was not used in this study [17]. These modifications led to a lower use of organic solvent to obtain the desired coumarins, without worsening the yields of the reactions.

4.2. Antifungal Activity of Coumarins against *C. albicans* and *S. cerevisiae* Strains

After preparing the coumarins, their ability to inhibit the growth of fluconazole-sensitive and fluconazole-resistant strains was evaluated. Only three compounds (ACS47, ACS50, and ACS54) presented antifungal activity, and all of them are substituted with a prenyl group. Their geranyl-substituted counterparts (ACS48, ACS51, and ACS52, respectively) did not display antifungal activity. ACS47 was active against the four strains tested, while ACS50 and ACS54 inhibited the growth of three strains and one strain, respectively. It may be observed that the insertion of methyl groups into the benzopyran moiety decreased the antifungal activity of the compounds, which may explain why ACS55 was not active against the yeasts, since it bears three methyl groups. ACS47 and ACS50 differ due to the presence of two methoxy substitutions in the latter. Results show that this substitution pattern does not influence the antifungal activity of the compound. Kurdelas et al. (2010) evaluated the antifungal activity of three prenylated coumarins with geranyl groups (auraptene, 5′-hydroxy auraptene, and 5′-oxo-auraptene) against yeasts and filamentous fungi and observed that none of the substances inhibited the growth of *C. albicans*, *C. tropicalis*, and *S. cerevisiae* [30], such as in this study. Jia et al. (2019) observed that coumarin (1,2-benzopyrone) exerts anti-*C. albicans* activity at 0.5 mg/mL–2 mg/mL by inducing apoptosis [31]. In this study, coumarins inhibited *C. albicans* and *S. cerevisiae* growth at 0.1 mg/mL. This action may be related to apoptosis induction, as seen in the unsubstituted coumarin, and the prenyl substitution could ease the interaction between the coumarins and plasma membrane, explaining the antifungal activity observed at a lower concentration [32].

4.3. Antifungal Activity of Coumarins Combined with Fluconazole against *S. cerevisiae* Strains

To evaluate whether the compounds inhibit *C. albicans* efflux pumps, hence precluding multidrug resistance and sensitizing the fungi to fluconazole, a screening assay using *S. cerevisiae* mutant strains was performed. Each strain heterologously expresses a specific MDR transporter, making it possible to comprehend exactly which efflux pumps the coumarins may affect. All coumarins, except for ACS55, improved the fluconazole antifungal activity against the *S. cerevisiae* strain that overexpresses CaMdr1p, an MFS transporter. This finding indicates that the compounds are able to inhibit this specific efflux pump, regardless of whether the coumarin possesses a prenyl or a geranyl substitution. ACS55, the only coumarin that did not affect CaMdr1p in this assay, bears three methyl groups that may impair its binding to the transporter. Moreover, data show that ACS47 and ACS50 also inhibit ABC transporters. These two coumarins carry a prenyl-substitution and are unsubstituted in the lactone ring. The presence of one methyl group is the only difference between ACS47, which inhibits CaCdr1p and CaCdr2p, as well as ACS54, which did not affect ABC transporters. In a previous study, it was observed that beta-lapachone impaired CgCdr2p ATPase activity, while beta-nor-lapachone did not have an inhibitory activity, and the difference between these two naphthoquinones is a methylene group [21]. Esposito et al. (2017) also observed that minor modifications in the structures of jatrophanes modified their inhibitory activity against CaCdr1p and CaMdr1p [33]. These data suggest that small increases in substitution may impact the inhibition of *Candida* spp. efflux transporters, which is important to optimize the design of new MDR inhibitors. The results of the screening assay were confirmed using a checkerboard. Eight out of the eleven tested combinations displayed synergism, and compounds ACS47 and ACS54 were the most active substances. Also, when comparing FICI values, it was not possible to determine a link between the presence of prenyl or geranyl substitutions and the inhibitory activity of the coumarins on the MFS transporter. To our knowledge, there are no reports on the ability of prenylated coumarins to inhibit *Candida* efflux pumps. However, the effects of geraniol have already been described. Singh et al. (2018) assessed the influence of the monoterpene on CaCdr1p and CaMdr1p, and observed that it affects only the ABC transporter. In silico analysis revealed that geraniol binds to CaCdr1p via hydrophobic interactions and H-bonds, while the control compound farnesol binds to the efflux pump only via hydrophobic interactions. Also, geraniol presented a lower binding energy than farnesol, indicating a higher affinity to the active site of CaCdr1p [34]. ACS47 and ACS50 are prenyl-substituted, and it may be hypothesized that the reduced size of the prenyl side chain, in comparison to a geranyl or a farnesyl chain, would increase the affinity of the coumarins to CaCdr1p, explaining the ability of these specific two compounds to inhibit ABC transporters. Aside from geraniol, the inhibitory activity of coumarin scopoletin on efflux pumps was also evaluated [35]. This substance is an unprenylated coumarin, presenting one hydroxyl and one methoxy substitution. Scopoletin exerted an antifungal effect on *C. tropicalis*, and at a sub-inhibitory concentration, promoted a four-fold decrease in fluconazole MIC, an effect related to the inhibition of efflux pumps. Although geraniol and scopoletin separately modulate *Candida* spp. efflux pumps, the geranylated coumarins tested in this study were unable to inhibit ABC transporters. One of the substances tested, ACS48 (also known as auraptene), induced the expression of P-glycoprotein in human intestinal cells [36]. Since this efflux pump is analogous to CaCdr1p and CaCdr2p, ACS48 and the other coumarins could induce the expression of *CDR1 and CDR2* instead of inhibiting the transporters, which may explain the lack of activity in these substances combined with fluconazole against *S. cerevisiae* mutant strains that overexpress ABC transporters.

4.4. Antifungal Activity of Coumarins Combined with Fluconazole against Strains of C. albicans

The *S. cerevisiae* mutant strains used in this study are important tools to assess the action of substances in individual efflux pumps. Nonetheless, to endorse a possible application of these compounds combined with fluconazole, it is essential to test them against pathogenic strains. Thus, fluconazole-resistant *C. albicans* clinical strains were also employed in this study. Data show that none of the compounds were synergistic with fluconazole against PRI strain. It is important to consider that FICI is a parameter that may not be strictly followed. In this study, FICI of 0.516 were obtained using ACS47 and ACS50 combined with fluconazole and, according to these results, the interactions are additive. However, the fact that no synergism was observed does not disqualify these substances as promising candidates to be used in clinics. Both coumarins reduced fluconazole MIC by 64 times, reinforcing their ability to inhibit efflux pumps. Although the other six coumarins can also inhibit CaMdr1p, indifferent interactions were detected against PRI strain. These compounds present geranyl groups and/or substitutions in the lactone ring, increasing their hydrophobicity. Since the plasma membranes from *S. cerevisiae* and *C. albicans* differ in composition, the diffusion of more hydrophobic substances across the membrane may be hindered in *C. albicans* [37]. Therefore, the impossibility of concentrating in the cytoplasm may justify the lack of activity of some compounds, even when they are able to inhibit efflux pumps. This could explain why ACS47 and ACS50, two prenyl-substituted coumarins, performed better than the other substances. ACS47 and ACS50 also improved the antifungal activity of fluconazole against 95-142. This *C. albicans* strain overexpresses two ABC transporters, CaCdr1p and CaCdr2p, and ACS47 is able to block both of them. Interestingly, ACS50 inhibits only CaCdr2p, but synergized with fluconazole against 95-142. This could be explained by a possible lower expression of *CDR1* in comparison to *CDR2*. In this scenario, the rate of fluconazole extrusion by CaCdr1p would not be sufficient, and the inhibition of CaCdr2p would promote fluconazole intracellular accumulation, allowing the antifungal agent to exert its activity. Moraes et al. (2020) observed that the naphthoquinone beta-lapachone exhibits the same inhibitory pattern of ACS50, i.e., the compound inhibits only CaCdr2p but displays synergism with fluconazole against the 95-142 strain [38].

4.5. Toxicity

Aside from being effective in treating candidiasis, alone and/or combined with fluconazole, the prenylated coumarins tested in this study must be safe for human use to be good candidates as antifungal agents. In this study, in vitro and in silico assays were conducted to verify the toxicity of the coumarins. The substances had a low hemolytic effect, which is essential to ensure the safety of a substance, since drugs reach the bloodstream after administration to the infection site. The lack of in vitro toxicity corroborates the findings of Maleki et al. (2019), who assessed the in vitro toxicity of O-prenylated coumarins against an HDF cell line and observed that ACS47, ACS48 and umbelliprenin did not affect its viability [39].

Although the absence of hemolysis is an important feature of any drug, it may be considered an initial test to assess toxicity [40]. Other cells constitute the human body, and a non-hemolytic drug may not be exempt from toxic effects. To better understand the toxicity profile of the coumarins, an in silico analysis was performed. None of the compounds showed mutagenic or tumorigenic effects. However, eight compounds were predicted to be irritant, and four compounds may have reproductive effects. Comparing the compounds, it is possible to observe that the presence of three methyl groups in the coumarin moiety induces irritant effects, regardless of the presence of a prenyl or a geranyl substitution. Among the coumarins with less substitutions, the prenylated coumarins do not have irritant effects, while their geranylated counterparts are toxic. On the other hand, all the compounds that bear a methyl group in the lactone ring cause reproductive effects, aside from the presence of a prenyl or geranyl substitution. The toxicity profiles predicted for ACS47 and ACS50 are similar to that observed for fluconazole, reinforcing a possible role of these two coumarins as new anti-*Candida* agents. The in silico analysis also provided LD_{50} for different routes of administration. The intraperitoneal and intravenous LD_{50} values for all coumarins were lower than those obtained for fluconazole. Contrarily, the coumarins presented higher oral and subcutaneous LD_{50} values than fluconazole. These results corroborate the lack of toxicity and encourage future studies with animals to test the in vivo activity of the coumarins against *C. albicans*.

5. Conclusions

In this study, we were able to improve the synthesis of altissimacoumarin D. Using a smaller amount of organic solvent, altissimacoumarin D and analogues were obtained in good yields and purity. ACS47, ACS50 and ACS54 presented antifungal activity alone, while all the coumarins (except for ACS55) inhibited at least one efflux pump. ACS48, ACS51, ACS52, ACS54 and ACS56 inhibited only CaMdr1p. No classic CaMdr1p inhibitors are available to be used as positive controls in experiments, and these coumarins could be used to fill this gap. ACS47 and ACS50 have synergistic activity when combined with fluconazole against the growth of *Candida albicans*. This effect is related to the inhibition of the efflux pumps CaCdr1p, CaCdr2p, and/or CaMdr1p. ACS47 inhibits all three transporters, which is important considering that the goal is to develop a new antifungal treatment since the substance is able to enhance fluconazole activity regardless of the efflux pump related to the resistance. Also, these compounds did not present toxicity in vitro and in silico. Therefore, ACS47 and ACS50 emerge as potential candidates for use as adjuvants in candidiasis therapy caused by resistant strains.

Supplementary Materials: The following supporting information can be downloaded at: https://www.mdpi.com/article/10.3390/jof9070758/s1. Figure S1. ^1H NMR spectra of ACS51; Figure S2. ^{13}C NMR spectra of ACS51; Figure S3. ^1H NMR spectra of ACS50; Figure S4. ^{13}C NMR spectra of ACS50; Figure S5. ^1H NMR spectra of ACS48; Figure S6. ^{13}C NMR spectra of ACS48; Figure S7. ^1H NMR spectra of ACS47; Figure S8. ^{13}C NMR spectra of ACS47; Figure S9. ^1H NMR spectra of ACS52; Figure S10. ^{13}C NMR spectra of ACS52; Figure S11. ^1H NMR spectra of ACS54; Figure S12. ^{13}C NMR spectra of ACS54; Figure S13. ^1H NMR spectra of ACS56; Figure S14. ^{13}C NMR spectra of ACS56; Figure S15. ^1H NMR spectra of ACS55; Figure S16. ^{13}C NMR spectra of ACS55.

Author Contributions: Conceptualization, A.G. and C.C.L.; methodology, A.C.S., D.C.d.M., D.C.d.C. and G.C.C.G.; data curation, A.C.S. and D.C.d.M.; writing—original draft preparation, A.C.S. and D.C.d.M.; writing—review and editing, A.F-P. and C.C.L.; visualization, A.G.; supervision, R.S.C.L., A.F.-P. and C.C.L.; project administration, C.C.L.; funding acquisition, R.S.C.L. and C.C.L. All authors have read and agreed to the published version of the manuscript.

Funding: This research was funded by Coordenação de Aperfeiçoamento de Pessoal de Nível Superior–Brazil (CAPES)–Finance Code 001.

Institutional Review Board Statement: Not applicable.

Informed Consent Statement: Not applicable.

Data Availability Statement: Not applicable.

Acknowledgments: We thank UFRJ for the facilities used in the elaboration of this work, and CNPq and FAPERJ for financial assistance.

Conflicts of Interest: The authors declare that they have no known competing financial interests or personal relationships that could have appeared to influence the work reported in this paper.

References

1. Soll, D.R.; Galask, R.; Schmid, J.; Hanna, C.; Mac, K.; Morrow, B. Genetic Dissimilarity of Commensal Strains of Candida Spp. Carried in Different Anatomical Locations of the Same Healthy Women. *J. Clin. Microbiol.* **1991**, *29*, 1702–1710. [CrossRef] [PubMed]
2. Brown, G.D.; Denning, D.W.; Gow, N.A.R.; Levitz, S.M.; Netea, M.G.; White, T.C. Hidden Killers: Human Fungal Infections. *Sci. Transl. Med.* **2012**, *4*, 165rv13. [CrossRef] [PubMed]
3. Lewis, R.E. Current Concepts in Antifungal Pharmacology. *Mayo Clin. Proc.* **2011**, *86*, 805–817. [CrossRef] [PubMed]
4. Franz, R.; Kelly, S.L.; Lamb, D.C.; Kelly, D.E.; Ruhnke, M.; Morschhäuser, J. Multiple Molecular Mechanisms Contribute to a Stepwise Development of Fluconazole Resistance in Clinical Candida Albicans Strains. *Antimicrob. Agents Chemother.* **1998**, *42*, 3065–3072. [CrossRef] [PubMed]
5. White, T.C.; Holleman, S.; Dy, F.; Mirels, L.F.; Stevens, D.A. Resistance Mechanisms in Clinical Isolates of *Candida Albicans. Agents Chemother.* **2002**, *46*, 1704–1713. [CrossRef]
6. Ter Beek, J.; Guskov, A.; Slotboom, D.J. Structural Diversity of ABC Transporters. *J. Gen. Physiol.* **2014**, *143*, 419–435. [CrossRef]
7. Kumari, S.; Kumar, M.; Khandelwal, N.K.; Kumari, P.; Varma, M.; Vishwakarma, P.; Shahi, G.; Sharma, S.; Lynn, A.M.; Prasad, R.; et al. ABC Transportome Inventory of Human Pathogenic Yeast Candida Glabrata: Phylogenetic and Expression Analysis. *PLoS ONE* **2018**, *13*, e0202993. [CrossRef]
8. Cannon, R.D.; Lamping, E.; Holmes, A.R.; Niimi, K.; Baret, P.V.; Keniya, M.V.; Tanabe, K.; Niimi, M.; Goffeau, A.; Monk, B.C. Efflux-Mediated Antifungal Drug Resistance. *Clin. Microbiol. Rev.* **2009**, *22*, 291–321. [CrossRef]
9. Li, X.; Li, Y.; Ma, S.; Zhao, Q.; Wu, J.; Duan, L.; Xie, Y.; Wang, S. Traditional Uses, Phytochemistry, and Pharmacology of Ailanthus Altissima (Mill.) Swingle Bark: A Comprehensive Review. *J. Ethnopharmacol.* **2021**, *275*, 114121. [CrossRef]
10. Qin, H.L.; Zhang, Z.W.; Ravindar, L.; Rakesh, K.P. Antibacterial Activities with the Structure-Activity Relationship of Coumarin Derivatives. *Eur. J. Med. Chem.* **2020**, *207*, 112832. [CrossRef]
11. Tan, N.; Yazıcı-Tütüniş, S.; Bilgin, M.; Tan, E.; Miski, M. Antibacterial Activities of Pyrenylated Coumarins from the Roots of Prangos Hulusii. *Molecules* **2017**, *22*, 1098. [CrossRef] [PubMed]
12. Fotso, G.W.; Ngameni, B.; Storr, T.E.; Ngadjui, B.T.; Mafu, S.; Stephenson, G.R. Synthesis of Novel Stilbene–Coumarin Derivatives and Antifungal Screening of Monotes Kerstingii-Specialized Metabolites against Fusarium Oxysporum. *Antibiotics* **2020**, *9*, 537. [CrossRef] [PubMed]
13. Iranshahi, M.; Arfa, P.; Ramezani, M.; Jaafari, M.R.; Sadeghian, H.; Bassarello, C.; Piacente, S.; Pizza, C. Sesquiterpene Coumarins from Ferula Szowitsiana and in Vitro Antileishmanial Activity of 7-Prenyloxycoumarins against Promastigotes. *Phytochemistry* **2007**, *68*, 554–561. [CrossRef] [PubMed]
14. Tayarani-Najaran, Z.; Tayarani-Najaran, N.; Eghbali, S. A Review of Auraptene as an Anticancer Agent. *Front. Pharmacol.* **2021**, *12*, 698352. [CrossRef] [PubMed]
15. Liu, Y.P.; Yan, G.; Xie, Y.T.; Lin, T.C.; Zhang, W.; Li, J.; Wu, Y.J.; Zhou, J.Y.; Fu, Y.H. Bioactive Prenylated Coumarins as Potential Anti-Inflammatory and Anti-HIV Agents from Clausena Lenis. *Bioorg. Chem.* **2020**, *97*, 103699. [CrossRef] [PubMed]
16. Dao, T.T.; Tran, T.L.; Kim, J.; Nguyen, P.H.; Lee, E.H.; Park, J.; Jang, I.S.; Oh, W.K. Terpenylated Coumarins as SIRT1 Activators Isolated from Ailanthus Altissima. *J. Nat. Prod.* **2012**, *75*, 1332–1338. [CrossRef] [PubMed]

17. Silva, A.C.; Benelkebir, H.; Lopes, R.S.C.; Lopes, C.C.; Ganesan, A. Total Synthesis of Altissimacoumarin D, a Small Molecule Sirtuin 1 Activator. *J. Braz. Chem. Soc.* **2018**, *29*, 1157–1161.
18. Van Schijndel, J.; Molendijk, D.; Canalle, L.A.; Rump, E.T.; Meuldijk, J. Temperature Dependent Green Synthesis of 3-Carboxycoumarins and 3,4-Unsubstituted Coumarins. *Curr. Org. Synth.* **2018**, *16*, 130–135. [CrossRef]
19. Lamping, E.; Monk, B.C.; Niimi, K.; Holmes, A.R.; Tsao, S.; Tanabe, K.; Niimi, M.; Uehara, Y.; Cannon, R.D. Characterization of Three Classes of Membrane Proteins Involved in Fungal Azole Resistance by Functional Hyperexpression in *Saccharomyces cerevisiae*. *Eukaryot. Cell* **2007**, *6*, 1150–1165. [CrossRef]
20. Pinto, A.C.C.; Rocha, D.A.S.; Moraes, D.C.D.; Junqueira, M.L.; Ferreira-Pereira, A. Candida Albicans Clinical Isolates from a Southwest Brazilian Tertiary Hospital Exhibit MFS-Mediated Azole Resistance Profile. *An. Acad. Bras. Cienc.* **2019**, *91*, e20180654. [CrossRef]
21. Clemente de Moraes, D.; do Carmo Freire Ribeiro Pinto, M.; Tenório Sousa Domingos, L.; do Valle Pereira Midlej, V.; Ferreira-Pereira, A. Effects of β-Lapachone and β-nor-Lapachone on Multidrug Efflux Transporters and Biofilms of Candida Glabrata. *Bioorg. Med. Chem.* **2022**, *63*, 116749. [CrossRef] [PubMed]
22. Niimi, K.; Harding, D.R.K.; Parshot, R.; King, A.; Lun, D.J.; Decottignies, A.; Niimi, M.; Lin, S.; Cannon, R.D.; Goffeau, A.; et al. Chemosensitization of Fluconazole Resistance in Saccharomyces Cerevisiae and Pathogenic Fungi by a D -Octapeptide Derivative. *Antimicrob. Agents Chemother.* **2004**, *48*, 1256–1271. [CrossRef] [PubMed]
23. de Moraes, D.C.; Cardoso, K.M.; Domingos, L.T.S.; do Carmo Freire Ribeiro Pinto, M.; Monteiro, R.Q.; Ferreira-Pereira, A. β-Lapachone Enhances the Antifungal Activity of Fluconazole against a Pdr5p-Mediated Resistant Saccharomyces Cerevisiae Strain. *Braz. J. Microbiol.* **2020**, *51*, 1051–1060. [CrossRef] [PubMed]
24. Lagunin, A.; Zakharov, A.; Filimonov, D.; Poroikov, V. QSAR Modelling of Rat Acute Toxicity on the Basis of PASS Prediction. *Mol. Inform.* **2011**, *30*, 241–250. [CrossRef] [PubMed]
25. Cho, S.K.; Jeong, M.; Jang, D.S.; Choi, J.H. Anti-Inflammatory Effects of Canthin-6-One Alkaloids from Ailanthus Altissima. *Planta Med.* **2018**, *84*, 527–535. [CrossRef]
26. Jeong, M.; Kim, H.M.; Ahn, J.H.; Lee, K.T.; Jang, D.S.; Choi, J.H. 9-Hydroxycanthin-6-One Isolated from Stem Bark of Ailanthus Altissima Induces Ovarian Cancer Cell Apoptosis and Inhibits the Activation of Tumor-Associated Macrophages. *Chem. Biol. Interact.* **2018**, *280*, 99–108. [CrossRef]
27. Shao, Q.; Li, T.Y.; Quan, W.; Wang, H.; Yan, K.Y.; Qiu, P.C.; Tang, H.F.; Lu, Y.Y. The Alkaloids with Neuroprotective Effect from the Root Bark of Ailanthus Altissima. *Nat. Prod. Res.* **2022**, 1–7. [CrossRef]
28. Li, Y.; Zhao, M.; Zhang, Z. Quantitative Proteomics Reveals the Antifungal Effect of Canthin-6-One Isolated from Ailanthus Altissima against Fusarium Oxysporum f. Sp. Cucumerinum in Vitro. *PLoS ONE* **2021**, *16*, e0250712. [CrossRef]
29. Annunziata, F.; Pinna, C.; Dallavalle, S.; Tamborini, L.; Pinto, A. An Overview of Coumarin as a Versatile and Readily Accessible Scaffold with Broad-Ranging Biological Activities. *Int. J. Mol. Sci.* **2020**, *21*, 4618. [CrossRef]
30. Kurdelas, R.R.; Lima, B.; Tapia, A.; Feresin, G.E.; Sierra, M.G.; Rodríguez, M.V.; Zacchino, S.; Enriz, R.D.; Freile, M.L. Antifungal Activity of Extracts and Prenylated Coumarins Isolated from Baccharis Darwinii Hook & Arn. (Asteraceae). *Molecules* **2010**, *15*, 4898–4907. [CrossRef]
31. Jia, C.; Zhang, J.; Yu, L.; Wang, C.; Yang, Y.; Rong, X.; Xu, K.; Chu, M. Antifungal Activity of Coumarin against Candida Albicans Is Related to Apoptosis. *Front. Cell. Infect. Microbiol.* **2019**, *9*, 445. [CrossRef] [PubMed]
32. Chen, X.; Mukwaya, E.; Wong, M.S.; Zhang, Y. A Systematic Review on Biological Activities of Prenylated Flavonoids. *Pharm. Biol.* **2014**, *52*, 655–660. [CrossRef] [PubMed]
33. Esposito, M.; Nim, S.; Nothias, L.F.; Gallard, J.F.; Rawal, M.K.; Costa, J.; Roussi, F.; Prasad, R.; Di Pietro, A.; Paolini, J.; et al. Evaluation of Jatrophane Esters from Euphorbia Spp. as Modulators of Candida Albicans Multidrug Transporters. *J. Nat. Prod.* **2017**, *80*, 479–487. [CrossRef] [PubMed]
34. Singh, S.; Fatima, Z.; Ahmad, K.; Hameed, S. Fungicidal Action of Geraniol against Candida Albicans Is Potentiated by Abrogated CaCdr1p Drug Efflux and Fluconazole Synergism. *PLoS ONE* **2018**, *13*, e0203079. [CrossRef]
35. Lemos, A.S.O.; Florêncio, J.R.; Pinto, N.C.C.; Campos, L.M.; Silva, T.P.; Grazul, R.M.; Pinto, P.F.; Tavares, G.D.; Scio, E.; Apolônio, A.C.M.; et al. Antifungal Activity of the Natural Coumarin Scopoletin Against Planktonic Cells and Biofilms from a Multidrug-Resistant Candida Tropicalis Strain. *Front. Microbiol.* **2020**, *11*, 1525. [CrossRef]
36. Nabekura, T.; Kawasaki, T.; Kato, Y.; Kawai, K.; Fiorito, S.; Epifano, F.; Genovese, S.; Uwai, Y. Citrus Auraptene Induces Drug Efflux Transporter P-Glycoprotein Expression in Human Intestinal Cells. *Food Funct.* **2020**, *11*, 5017–5023. [CrossRef]
37. Mukhopadhyay, K.; Kohli, A.; Prasad, R. Drug Susceptibilities of Yeast Cells Are Affected by Membrane Lipid Composition. *Antimicrob. Agents Chemother.* **2002**, *46*, 3695–3705. [CrossRef]
38. Moraes, D.C.; Reis de Sá, L.F.; Domingos, L.T.S.; Pinto, M.d.C.F.R.; Soares, R.M.d.A.; Ferreira-Pereira, A. Synergistic Interactions between β-Lapachone and Fluconazole in the Inhibition of CaCdr2p and CaMdr1p in Candida Albicans. *Rev. Iberoam. Micol.* **2020**, *37*, 104–106. [CrossRef]

39. Maleki, E.H.; Bahrami, A.R.; Sadeghian, H.; Matin, M.M. Discovering the Structure–Activity Relationships of Different O-Prenylated Coumarin Derivatives as Effective Anticancer Agents in Human Cervical Cancer Cells. *Toxicol. Vitr.* **2020**, *63*, 104745. [CrossRef]
40. Greco, I.; Molchanova, N.; Holmedal, E.; Jenssen, H.; Hummel, B.D.; Watts, J.L.; Håkansson, J.; Hansen, P.R.; Svenson, J. Correlation between Hemolytic Activity, Cytotoxicity and Systemic in Vivo Toxicity of Synthetic Antimicrobial Peptides. *Sci. Rep.* **2020**, *10*, 13206. [CrossRef]

Journal of
Fungi

MDPI

Article

The Transcriptome Response to Azole Compounds in *Aspergillus fumigatus* Shows Differential Gene Expression across Pathways Essential for Azole Resistance and Cell Survival

Margriet W. J. Hokken [1,2,*], Jordy P. M. Coolen [1,2], Hilbert Steenbreker [1], Jan Zoll [1,2], Tim J. H. Baltussen [1,2], Paul E. Verweij [1,2] and Willem J. G. Melchers [1,2]

[1] Department of Medical Microbiology, Radboud Institute for Molecular Life Sciences, Radboud University Medical Center, 6500 HB Nijmegen, The Netherlands; tim.baltussen@radboudumc.nl (T.J.H.B.)
[2] Center of Expertise in Mycology Radboudumc/CWZ, 6500 HB Nijmegen, The Netherlands
* Correspondence: margriet.hokken@radboudumc.nl

Abstract: The opportunistic pathogen *Aspergillus fumigatus* is found on all continents and thrives in soil and agricultural environments. Its ability to readily adapt to novel environments and to produce billions of spores led to the spread of azole-resistant *A. fumigatus* across the globe, posing a threat to many immunocompromised patients, including critically ill patients with severe influenza or COVID-19. In our study, we sought to compare the adaptational response to azoles from *A. fumigatus* isolates that differ in azole susceptibility and genetic background. To gain more insight into how short-term adaptation to stressful azole compounds is managed through gene expression, we conducted an RNA-sequencing study on the response of *A. fumigatus* to itraconazole and the newest clinically approved azole, isavuconazole. We observed many similarities in ergosterol biosynthesis up-regulation across isolates, with the exception of the pan-azole-resistant isolate, which showed very little differential regulation in comparison to other isolates. Additionally, we found differential regulation of membrane efflux transporters, secondary metabolites, iron metabolism, and various stress response and cell signaling mechanisms.

Keywords: *Aspergillus fumigatus*; azole resistance; isavuconazole; itraconazole; RNA-seq; transcriptome

1. Introduction

The ascomycete *Aspergillus fumigatus* is a pathogenic fungus that is found across all continents [1,2]. *A. fumigatus* spreads through airborne conidia, which can cause a wide spectrum of diseases in immunocompromised patients depending on the severity of the underlying immunodeficiencies that can result from genetic disorders in both innate and adaptive immune responses, underlying diseases, chemotherapy, or post-transplant immunosuppression [3]. Furthermore, *A. fumigatus* can produce a variety of virulence factors such as gliotoxin, which influence pathogenesis [4]. Lastly, it has been proposed that the pulmonary microbiome influences the severity of infections caused by *A. fumigatus* [5,6].

Additionally, *Aspergillus* infections have been observed among critically ill influenza and COVID-19 patients in the intensive care unit (ICU) [7,8]. Its potential to easily spread across great distances, together with the ability to adapt to novel challenging environments, makes *A. fumigatus* a difficult-to-treat opportunistic pathogen [9,10]. Clinicians remain limited in treatment options; the azoles remain the first choice in the treatment and management of *Aspergillus* diseases. However, the efficacy of the azole class is threatened by the emergence of azole resistance [11].

Azole resistance can develop in patients with a history of azole treatment; however, many resistant *A. fumigatus* isolates are likely to have originated in the environment due to the use of agricultural azole derivatives [1,12–14]. Azole class compounds inhibit the sterol 14α-demethylase Cyp51A, an essential enzyme in ergosterol biosynthesis [15]. The majority of resistant isolates harbor a TR_{34}/L98H mutation on the *cyp51A* locus, which is highly likely to have an environmental origin due to the use of agricultural azole derivatives and subsequent natural selection of cross-resistant isolates [13].

A. fumigatus is able to readily adapt to new and stressful environments to maintain growth and reproduction [16–18]. Many mutations that enable the fungus to overcome azole pressure have been elucidated in the last decade. These include a wide spectrum of mutations in the *cyp51A* coding area and promotor region and mutations in regulatory genes involved in ergosterol biosynthesis [19–23]. However, not all resistant isolates harbor mutations in *cyp51A* or regulatory genes involved in ergosterol biosynthesis, and isolates can have increased MICs for azole compounds that are not explained by the mutations present in the isolate [24,25]. It has been suggested that several cellular processes ultimately contribute to the resistance profile of an isolate, including increased drug efflux pump activity, mitochondrial dysfunction, and activation of environmental stress response pathways [26–29].

As many of these molecular mechanisms undergo differential regulation of genes involved, transcriptome analysis can give insight into activated mechanisms upon adaptation to azole compounds and shed light on how *A. fumigatus* changes its physiology in order to survive. As individuals can be colonized with susceptible and resistant *A. fumigatus* conidia, we exposed azole-susceptible and resistant isolates to sub-inhibitory concentrations of the azole compounds itraconazole and isavuconazole to monitor changes in gene expression after 48h of azole exposure.

This research shows that several molecular mechanisms are activated upon azole addition to the medium. These findings can shed light on the genetic regulatory networks involved in cellular adaptation to azole compounds, which can enable us to design new antifungal therapies in the future.

2. Results

To investigate the transcriptome during exponential growth under antifungal pressure, we conducted an in vitro assay in which we exposed fresh conidia of *A. fumigatus* isolates to sublethal concentrations of itraconazole (ICZ) and isavuconazole (ISA) (Table 1).

Table 1. Isolates used in this study.

	MIC												TREATED WITH	
ISOLATE	ICZ	VOR	POS	ISA	Cyp51A Genotype	STR3A	STR3B	STR3C	STR4A	STR4B	STR4C	Reference	ICZ (IC50)	ISA (IC50)
V147(S)	0.5	1	0.25	1	WT	23	6	14	5	4	4	This study	0.15 μg/mL	0.10 μg/mL
V162(R)	>16	4	0.25	8	TR34/L98H	52	9	11	7	6	11	This study	8 μg/mL	1.98 μg/mL
V181(R)	>16	>16	1	>16	$TR46^3$/Y121F/M172I/T289A/G448S	26	8	12	6	7	7	[30]	8 μg/mL	16 μg/mL

The isolates used in this study are listed in Table 1. The mycelial morphology of these isolates grown on MM is shown in Figure S1. STR*af* data show that these isolates are indeed of a separate genotype. These genomic differences underline that the isolates tested in this study have likely originated from a different environment, which supports the assumption that the transcriptome alterations are at least suspected to be different to some extent between these isolates due to genetic differences.

These isolates are either susceptible, resistant to ICZ, or pan-azole-resistant. Mutations on the *cyp51A* locus explain the higher MICs in the latter two isolates, as they contain the well-known TR34/L98H mutation that confers resistance to ICZ, voriconazole (VCZ), and isavuconazole (ISA) and a $TR46^3$/Y121F/T289A mutation, which confers resistance to all clinical azoles [30]. The V181(R^{PAN}) isolate was originally isolated from a compost heap in The Netherlands [30]. Due to these characteristics, we named these isolates V147(S), V162(R), and V181(R^{PAN}).

The transcriptomic response of three strains was compared by inoculating each isolate with IC50 concentrations of ICZ or ISA, as determined by dose–response curves. These isolates were compared to isolates treated with an equal amount of DMSO.

Additionally, we tested the effect of DMSO on untreated samples of susceptible isolate V147(S). All reported log2 fold-change (log2FC) values can be found in Table S3 containing all DEGs after the treatment of these isolates (Table S3).

2.1. Exploratory Analysis of DEGs

On average, 18,420,393 uniquely mapped sequence reads were obtained per isolate, ranging from 6,235,921 to 25,617,320 reads per individual sample, corresponding to 32× and 177× genome coverage, respectively. After genome alignment, an average unique mapping percentage of 87.1% to reference genome Af293 was found [31] (Table S2).

Heatmap visualization of all differentially expressed genes (DEGs) shows strain-specific transcriptional differences after treatment with ICZ and ISA (Figure 1A). Principle component analysis (PCA) revealed that the transcriptome is highly variable between isolates and that the biological replicates clustered together for each isolate and treatment (Figure 1C). Susceptible isolate V147(S) showed a clear treatment effect for both ICZ and ISA, whereas resistant isolate V162(R) only showed a predominant treatment effect for ISA. Pan-azole-resistant isolate V181(R^{PAN}) did not show great variance between treatments and control samples. After ICZ treatment of the isolates, we identified 3102 and 873 DEGs for isolates V147(S) and V162(R), respectively, from a total of 9840 genes (Genome inventory, AF293 genome, AspGD, 1 April 2020) compared to untreated isolates. No DEGs were found for V181(R^{PAN}) after ICZ treatment. After ISA treatment of the isolates, we identified 1883, 1795, and 326 DEGs for isolates V147(S), V162(R), and V181(R^{PAN}), respectively (Figure 1B). A volcano plot for each time point was created to assess the distribution of the log2FC values relative to the corrected *p*-value (Figure 1E). To investigate if the organic solvent DMSO had any influence on gene expression, we treated susceptible isolate V147(S) with the same volume of DMSO and found 21 DEGs after treatment. The genes that were found were filtered from the list of DEGs retrieved after azole addition to the medium (Table S3). As these results were only available for isolate V147(S), it should be noted that this list might not represent the genetic diversity of all isolates and might not be exhaustive, as this was not compared for isolates V162(R) and V181(R^{PAN}).

Lastly, to obtain a better view on which genes are identically regulated in isolates V147(S) and V162(R) during both treatments, we overlapped all DEGs that were similarly regulated during both treatments in a Venn diagram. We found 63 up-regulated core DEGs and 62 down-regulated core DEGs, but the vast majority of DEGs in V147(S) were unique for that isolate (Figure 1D). No significantly enriched FunCat categories or GO terms were found for these gene sets. From all 125 shared core DEGs, only 9 are known and characterized. Among the up-regulated genes for V147(S) and V162(R) were efflux transporter *cdr1B* (Afu1g1433) and thaumatin domain protein *calB* (Afu8g01710), a paralogue of *calA* (Afu3g09690), a cell wall protein that mediates interaction with human integrin β1 and is required for normal virulence [32]. Among the down-regulated genes are siderophore biosynthesis genes SidA (Afu2g07680), and two other siderophore biosynthesis genes, (Afu3g03390) and (Afu3g03400). The other down-regulated core DEGs comprise genes involved in diverse cellular processes such as host–pathogen interaction, cell signaling, and ribosome biosynthesis. All shared DEGs are listed in Table S4.

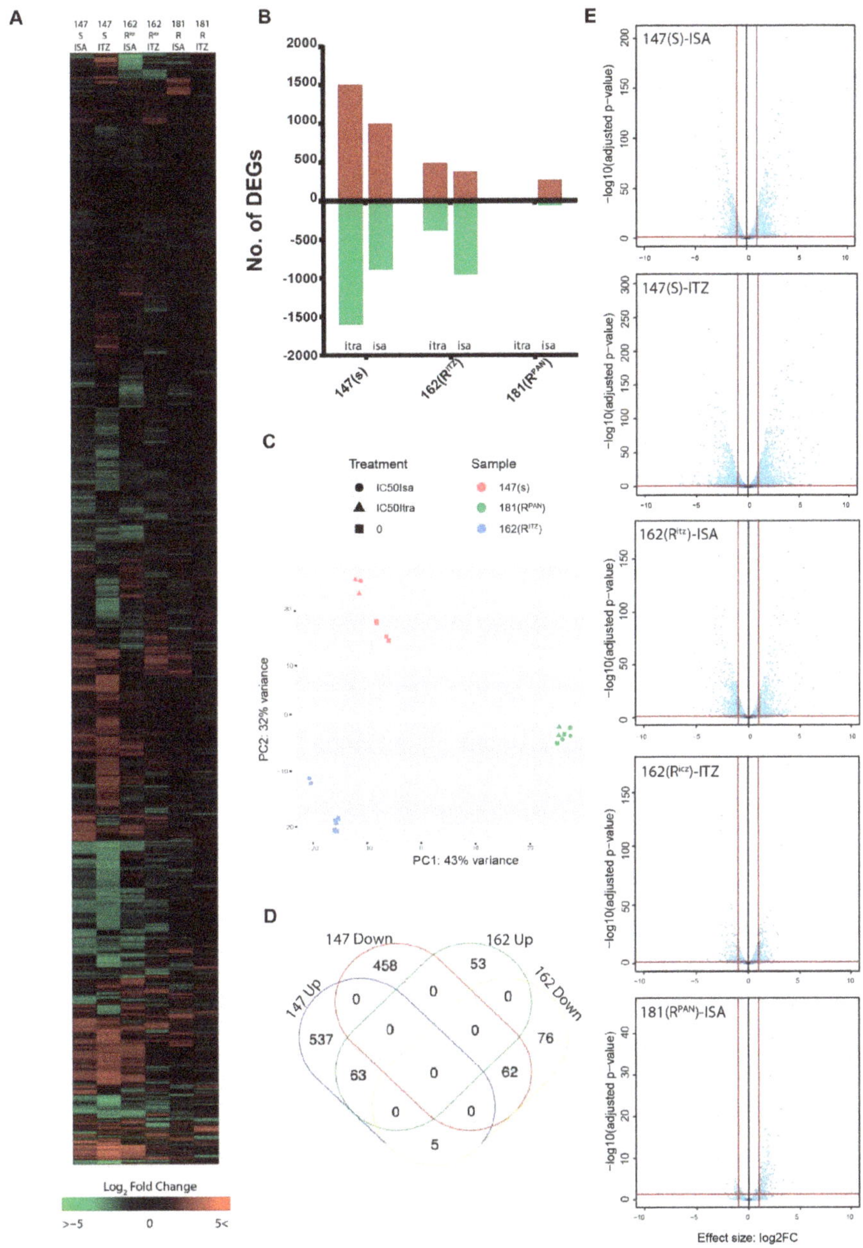

Figure 1. Exploratory results. (**A**). Heatmap of the total transcriptome after hierarchical clustering. (**B**). The total number of DEGs found in each isolate after treatment with ICZ or ISA. (**C**). Principal component analysis shows clustering of biological replicates. The effect of azole treatment on the transcriptome has less influence than the isolate background. (**D**). Overlap of similarly regulated DEGs of isolates V147(S) and V162(R) by Venn analysis. A total of 63 up-regulated core DEGs and 62 down-regulated core DEGs were found. (**E**). Volcano plots of each isolate and treatment. Log2FC values are plotted on the X-axis, whereas the −log10 (adjusted *p*-value) is plotted on the Y-axis. Please note that the Y-axis varies in maximum value when comparing volcano plots.

Remarkably, when we compared the 268 up-regulated DEGs and 56 down-regulated DEGs from pan-azole-resistant isolate V181(R^{PAN}) treated with ISA with these core DEGs, only 12 genes were shared between these isolates, underlining a strong difference in transcriptome adaptation between isolates. Among these genes were a majority of uncharacterized genes and ABC transporters Cdr1B (Afu1g14330) and AbcA (Afu2g15130).

2.2. Most DEGs after Azole Treatment

We analyzed the 50 DEGs with the highest positive or negative log2FC per isolate and treatment to assess the individual response of each isolate to different azole compounds (Table 2). The transcript encoding for integral membrane protein Afu2g17930 was the most up-regulated gene with a log2FC of 10.17 in V147(S)-ICZ, 5.37 in V147(S)-ISA, and 5.07 in V162(R) -ISA. This gene is known to be up-regulated upon acute ER stress [33]. It is predicted to play a role in fatty acid hydroxylation and oxidation–reduction processes (AspGD; Prediction of Gene Ontology (GO) annotations based on protein characteristics, 2011) and was also found up-regulated in voriconazole-treated isolates [34]. Several RTA1 domain proteins were present amongst the most up-regulated DEGs, of which Afu6g14140 was the most up-regulated in all isolates. Not much is known about this family of transmembrane proteins in *A. fumigatus*, but their homologs present in *Candida albicans* are also induced by azole compounds and involved in azole resistance through regulation via calcineurin signaling [35,36]. The ABC transporter AbcE (Afu7g00480) was found the most up-regulated in the pan-azole-resistant isolate V181(R^{PAN}) with log2FC 4.31 and was moderately up-regulated in isolates V147(S)-ICZ, V147(S)-ISA, and V162(R)-ICZ. Furthermore, we found strong up-regulation of a putative fungal adhesin (Afu2g10130) in V147(S)-ICZ with log2FC 7.3, which was also up-regulated in V147(S)-ISA and V162(R)-ISA. We also found up-regulation of the 'Aspf4-like' putative allergen (Afu3g00710) in isolates V162(R)-ICZ and V162(R)-ISA, with log2FC 2.14 and log2FC 3.89, respectively. Additionally, the major allergen Aspf1 (Afu5g02330) was found strongly up-regulated in isolate V147(S)-ICZ with log2FC 5.72 and was moderately up-regulated in V147(S)-ISA and V162(R)-ISA. Several genes present in secondary metabolite gene clusters were found among the most up-regulated genes including *hasC* and *nscA.* Furthermore, genes involved in siderophore biosynthesis and transport, *sidG* (Afu3g03650) and *mirB* (Afu3g03640), are amongst the most down-regulated in isolates V147(S)-ISA, V162(R)-ISA, and V162(R)-ICZ.

Table 2. List of the 50 DEGs with the highest and lowest log2FC for each isolate.

		V147(S)-ICZ	
		Up-regulated	
2logFC	gene	Description	Name
10,18	Afu2g17930	Integral membrane protein	N/A
9,15	Afu3g12910	O-methyltransferase GliM-like; putative	hasC
8,26	Afu8g06070	Putative uncharacterized protein	N/A
7,65	Afu4g02720	GPI anchored glycosyl hydrolase; putative	N/A
7,29	Afu3g12920	Nonribosomal peptide synthetase 5	nrps5
7,26	Afu2g10130	Adhesin; putative	N/A
7,25	Afu3g12900	MFS transporter; putative	hasB
7,19	Afu3g12930	7-dimethylallyltryptophan synthase	hasE
7,17	Afu3g12960	Cytochrome P450 monooxigenase GliC-like; putative	hasH
6,97	Afu7g04550	Serine/threonine protein kinase; putative	N/A
6,82	Afu6g11850	Putative uncharacterized protein	N/A
6,62	Afu7g04570	Na/K ATPase alpha 1 subunit; putative	N/A
6,45	Afu3g01500	Integral membrane protein	N/A
6,41	Afu4g01260	Mitochondrial chaperone ATPase (Bcs1); putative	bcs1
6,40	Afu3g01960	Putative uncharacterized protein	N/A
6,32	Afu8g00620	Dimethylallyl tryptophan synthase; putative	cdpNPT
6,24	Afu6g14140	RTA1 domain protein; putative	N/A

Table 2. *Cont.*

5,81	Afu4g01290	Glycosyl hydrolase family 75 chitosanase; hydrolyzes beta-1,4-glycosidic linkage of chitosan; repressed by exposure to artemisinin	csnB
5,72	Afu5g02330	Ribonuclease mitogillin	aspf1
5,72	Afu2g17810	RTA1 domain protein; putative	N/A
5,55	Afu3g03470	Putative uncharacterized protein	N/A
5,52	Afu7g04560	Putative uncharacterized protein	N/A
5,47	Afu2g16540	C2H2 finger domain protein; putative	N/A
5,42	Afu1g01175	Putative uncharacterized protein	N/A
5,25	Afu4g14030	Putative uncharacterized protein	N/A
		Down-regulated	
-3,67	Afu8g05940	Putative uncharacterized protein	N/A
-3,71	Afu3g02800	Lipase/esterase; putative	N/A
-3,72	Afu6g04360	ABC drug exporter AtrF	atrF
-3,74	Afu2g00460	Short-chain dehydrogenase/reductase family protein; putative	N/A
-3,78	Afu4g01560	MFS myo-inositol transporter; putative	N/A
-3,79	Afu2g00550	Putative uncharacterized protein	N/A
-3,81	Afu6g07820	Integral membrane protein; putative	N/A
-3,81	Afu3g14940	Elastase inhibitor AFUEI	aeiA
-3,87	Afu1g03352	Alpha-1;3-glucanase/mutanase; putative	N/A
-3,89	Afu6g00730	Putative uncharacterized protein	N/A
-4,03	Afu4g13570	Thiol methyltransferase; putative	N/A
-4,09	Afu1g17180	Pyridine nucleotide-disulphide oxidoreductase AMID-like; putative	N/A
-4,31	Afu7g00700	Aldo-keto reductase (AKR13); puatative	akr13
-4,32	Afu2g01590	Non-classical export protein Nce102; putative	nce102
-4,64	Afu6g03350	GNAT family N-acetyltransferase; putative	N/A
-4,67	Afu7g01040	Cytidine deaminase; putative	N/A
-4,75	Afu4g00450	Putative uncharacterized protein	N/A
-4,86	Afu3g13620	Cupin domain protein	N/A
-5,15	Afu8g06760	MFS transporter; putative	N/A
-5,15	Afu8g01290	Putative uncharacterized protein	N/A
-5,25	Afu1g16440	ABC ATPase; putative	N/A
-5,41	Afu2g05290	Isoflavone reductase family protein	N/A
-5,59	Afu5g01030	Glyceraldehyde-3-phosphate dehydrogenase	N/A
-6,32	Afu8g01660	Putative uncharacterized protein	N/A
-6,43	Afu6g03190	Putative uncharacterized protein	N/A
		V147(S)-ISA	
		Up-regulated	
2logFC	gene	Description	
5,74	Afu1g01175	Putative uncharacterized protein	N/A
5,36	Afu2g17930	Integral membrane protein	N/A
5,20	Afu7g00160	Polyketide synthase; putative	nscA
4,99	Afu7g00120	Metallo-beta-lactamase domain protein	nscB
4,80	Afu7g00170	Dimethylallyl tryptophan synthase GliD1	nscD
4,54	Afu7g00180	NAD dependent epimerase/dehydratase; putative	nscE
4,46	Afu7g00200	Putative uncharacterized protein	N/A
4,36	Afu4g01260	Mitochondrial chaperone ATPase (Bcs1); putative	bcs1
4,16	Afu3g01960	Putative uncharacterized protein	N/A
4,13	Afu6g14140	RTA1 domain protein; putative	N/A
4,06	Afu8g00620	Dimethylallyl tryptophan synthase; putative	cdpNPT
4,01	Afu8g06640	UbiE/COQ5 methyltransferase; putative	N/A
3,98	Afu4g01290	Glycosyl hydrolase family 75 chitosanase; hydrolyzes beta-1,4-glycosidic linkage of chitosan; repressed by exposure to artemisinin	csnB
3,95	Afu6g03680	Putative uncharacterized protein	N/A
3,93	Afu7g04550	Serine/threonine protein kinase; putative	N/A
3,80	Afu5g01230	RTA1 domain protein; putative	N/A
3,71	Afu7g04570	Na/K ATPase alpha 1 subunit; putative	N/A
3,63	Afu7g04560	Putative uncharacterized protein	N/A

Table 2. *Cont.*

3,61	Afu3g01500	Integral membrane protein	N/A
3,56	Afu8g02070	Glycosyl transferase; putative	N/A
3,46	Afu4g02720	GPI anchored glycosyl hydrolase; putative	N/A
3,38	Afu4g09550	Putative uncharacterized protein	N/A
3,36	Afu1g12420	Putative uncharacterized protein	N/A
3,33	Afu4g03630	Sterol 24-c-methyltransferase; putative	erg6
3,25	Afu3g10370	Putative uncharacterized protein	N/A
		Down-regulated	
-2,60	Afu2g00570	GNAT family acetyltransferase; putative	N/A
-2,61	Afu2g17850	3-beta hydroxysteroid dehydrogenase/isomerase; putative	N/A
-2,71	Afu1g17270	FRE family ferric-chelate reductase; putative	fre2
-2,73	Afu5g11250	Polyglutamate biosynthesis protein; putative	N/A
-2,74	Afu6g00750	Pyruvate decarboxylase; putative	pdcB
-2,75	Afu4g12740	tRNA (Adenine-N(1)-)-methyltransferase	N/A
-2,77	Afu6g03350	GNAT family N-acetyltransferase; putative	N/A
-2,77	Afu3g03650	GNAT family acetyltransferase; putative	sidG
-2,79	Afu3g03420	Nonribosomal peptide synthetase 4	sidD
-2,80	Afu5g13970	WD domain; G-beta repeat protein	N/A
-2,83	Afu6g00760	Glutathione S-transferase; putative	N/A
-2,89	Afu3g15050	Flavin-binding monooxygenase; putative	N/A
-2,91	Afu1g17170	TfdA family taurine dioxygenase; putative	N/A
-2,94	Afu3g03660	Siderophore esterase IroE-like; putative	estB
-2,98	Afu3g15055	Putative uncharacterized protein	N/A
-3,03	Afu4g00450	Putative uncharacterized protein	N/A
-3,26	Afu2g00580	Putative uncharacterized protein	N/A
-3,28	Afu2g18030	Catalase	fgaCat
-3,37	Afu1g03352	Alpha-1;3-glucanase/mutanase; putative	N/A
-3,39	Afu3g14940	Elastase inhibitor AFUEI	aeiA
-3,89	Afu3g03640	MFS siderochrome iron transporter MirB	mirB
-3,96	Afu3g15010	Putative uncharacterized protein	N/A
-4,39	Afu6g03190	Putative uncharacterized protein	N/A
-4,87	Afu3g03010	Phosphate-repressible Na+/phosphate cotransporter Pho89; putative	Pho89
-5,60	Afu5g01030	Glyceraldehyde-3-phosphate dehydrogenase	N/A
		V162(R)-ICZ	
		Up-Regulated	
2logFC	gene	Description	
3,89	Afu4g03630	Sterol 24-c-methyltransferase; putative	erg6
3,43	Afu7g04560	Putative uncharacterized protein	N/A
3,24	Afu2g15130	ABC multidrug transporter; putative	abcA
2,86	Afu2g08680	Putative uncharacterized protein	N/A
2,85	Afu6g14140	RTA1 domain protein; putative	N/A
2,75	Afu3g11880	Putative uncharacterized protein	N/A
2,48	Afu2g15140	MFS drug transporter; putative	N/A
2,29	Afu8g06640	UbiE/COQ5 methyltransferase; putative	N/A
2,26	Afu2g12820	Polyketide synthase; putative	N/A
2,25	Afu1g17470	High affinity nitrate transporter NrtB	nrtB
2,18	Afu1g06480	Putative uncharacterized protein	N/A
2,18	Afu1g03270	Putative uncharacterized protein	N/A
2,15	Afu3g00710	Allergen Asp F4-like; putative	N/A
2,13	Afu6g09960	Putative uncharacterized protein	N/A
2,12	Afu7g04880	Sterol glucosyltransferase; putative	N/A
2,03	Afu6g11720	Putative uncharacterized protein	N/A
1,98	Afu7g00910	OPT peptide transporter Mtd1; putative	optH
1,97	Afu1g06470	Neutral/alkaline nonlysosomal ceramidase; putative	N/A
1,94	Afu1g11360	Aldehyde reductase II	N/A
1,93	Afu2g00240	Putative uncharacterized protein	N/A
1,91	Afu3g09240	CAIB/BAIF family enzyme	N/A

Table 2. *Cont.*

1,89	Afu2g00410	Amidase family protein	N/A
1,88	Afu4g01410	Putative uncharacterized protein	N/A
1,86	Afu3g07860	Glycosyl transferase; putative	gtb3
1,85	Afu8g01010	Thermophilic desulfurizing enzyme family protein	N/A
		Down-Regulated	
-1,73	Afu5g00580	Putative uncharacterized protein	N/A
-1,74	Afu6g03190	Putative uncharacterized protein	N/A
-1,76	Afu7g06920	NmrA family transcriptional regulator; putative	N/A
-1,76	Afu3g12255	Putative uncharacterized protein	N/A
-1,77	Afu1g10380	Nonribosomal peptide synthetase 1	nrps1
-1,86	Afu8g05520	Aldehyde dehydrogenase family protein	N/A
-1,87	Afu7g06350	Sodium/phosphate symporter; putative	phoE
-1,88	Afu4g01350	Putative uncharacterized protein	gprK
-1,95	Afu4g00440	Short chain dehydrogenase; putative	N/A
-2,06	Afu2g17900	Putative uncharacterized protein	N/A
-2,07	Afu3g03640	MFS siderochrome iron transporter MirB	MirB
-2,07	Afu7g06150	Endoglucanase; putative	N/A
-2,14	Afu5g01620	Extracellular proline-rich protein	N/A
-2,14	Afu4g01360	MFS transporter of unkown specificity	N/A
-2,16	Afu1g10390	ABC multidrug transporter; putative	abcB
-2,20	Afu7g04920	Putative uncharacterized protein	N/A
-2,22	Afu2g00580	Putative uncharacterized protein	N/A
-2,25	Afu1g04130	FG-GAP repeat protein; putative	N/A
-2,32	Afu5g14740	Fucose-specific lectin FleA	fleA
-2,33	Afu1g03352	Alpha-1;3-glucanase/mutanase; putative	N/A
-2,53	Afu6g03560	IgE-binding protein; putative	N/A
-2,56	Afu3g14940	Elastase inhibitor AFUEI	aeiA
-2,68	Afu6g01870	Putative uncharacterized protein	N/A
-2,70	Afu6g11840	Sodium bile acid symporter family protein	N/A
-2,70	Afu6g00430	IgE-binding protein	N/A
-2,76	Afu8g02000	Sorbitol/xylitol dehydrogenase; putative	N/A
		V162(R)-ISA)	
		Up-Regulated	
2logFC	gene	Description	
8,95	Afu3g12910	O-methyltransferase GliM-like; putative	hasC
6,59	Afu3g01960	Putative uncharacterized protein	N/A
6,58	Afu6g11850	Putative uncharacterized protein	N/A
6,25	Afu4g02720	GPI anchored glycosyl hydrolase; putative	N/A
6,14	Afu1g01175	Putative uncharacterized protein	N/A
5,41	Afu8g06070	Putative uncharacterized protein	N/A
5,28	Afu7g04570	Na/K ATPase alpha 1 subunit; putative	N/A
5,12	Afu7g04550	Serine/threonine protein kinase; putative	N/A
5,12	Afu4g01290	Glycosyl hydrolase family 75 chitosanase; hydrolyzes beta-1,4-glycosidic linkage of chitosan; repressed by exposure to artemisinin	csnB
5,07	Afu2g17930	Integral membrane protein	N/A
4,95	Afu3g12920	Nonribosomal peptide synthetase 5	nrps5
4,86	Afu3g12960	Cytochrome P450 monooxigenase GliC-like; putative	hasH
4,85	Afu8g04920	LEA domain protein	N/A
4,76	Afu6g14140	RTA1 domain protein; putative	N/A
4,61	Afu6g03490	Phenol 2-monooxygenase; putative	fmpF
4,55	Afu5g08730	Putative uncharacterized protein	N/A
4,50	Afu4g00280	Putative uncharacterized protein	N/A
4,46	Afu6g11840	Sodium bile acid symporter family protein	N/A
4,34	Afu4g01260	Mitochondrial chaperone ATPase (Bcs1); putative	bcs1
4,06	Afu3g10370	Putative uncharacterized protein	N/A
4,00	Afu5g01230	RTA1 domain protein; putative	N/A
3,98	Afu3g01500	Integral membrane protein	N/A
3,93	Afu3g12930	7-dimethylallyltryptophan synthase	hasE

Table 2. *Cont.*

3,90	Afu8g00610	Cell surface protein Mas1; putative	mas1
3,90	Afu3g00710	Allergen Asp F4-like; putative	N/A
		Down-Regulated	
-2,90	Afu3g15100	Integral membrane protein; putative	N/A
-3,02	Afu7g00910	OPT peptide transporter Mtd1; putative	Mtd1
-3,07	Afu8g01850	Phosphate-repressible phosphate permease; putative	N/A
-3,09	Afu4g00450	Putative uncharacterized protein	N/A
-3,11	Afu1g15180	Putative uncharacterized protein	N/A
-3,11	Afu3g03420	Nonribosomal peptide synthetase 4	sidD
-3,12	Afu6g13930	LovB-like polyketide synthase; putative	pyr2
-3,15	Afu6g03190	Putative uncharacterized protein	N/A
-3,16	Afu4g01560	MFS myo-inositol transporter; putative	N/A
-3,17	Afu4g00830	MFS peptide transporter; putative	N/A
-3,23	Afu7g00420	Putative uncharacterized protein	N/A
-3,25	Afu5g02860	Integral membrane protein; putative	N/A
-3,28	Afu5g02850	Putative uncharacterized protein	N/A
-3,29	Afu8g06090	Amino acid permease; putative	N/A
-3,34	Afu6g01870	Putative uncharacterized protein	N/A
-3,39	Afu1g04130	FG-GAP repeat protein; putative	N/A
-3,42	Afu8g06100	Integral membrane protein	N/A
-3,45	Afu1g17580	Xenobiotic compound monooxygenase; DszA family; putative	N/A
-3,47	Afu7g06150	Endoglucanase; putative	N/A
-3,73	Afu7g06140	Probable beta-glucosidase L	exg13
-4,11	Afu4g13850	TfdA family oxidoreductase; putative	N/A
-4,13	Afu3g03650	GNAT family acetyltransferase; putative	sidG
-4,53	Afu1g03352	Alpha-1;3-glucanase/mutanase; putative	N/A
-4,70	Afu3g03640	MFS siderochrome iron transporter MirB	MirB
-5,18	Afu7g00440	GABA permease; putative	N/A
		V181(R^{PAN})-ISA	
		Up-Regulated	
2logFC	gene	Description	
4,31	Afu7g00480	ABC multidrug transporter; putative	abcE
3,21	Afu3g03530	Nitroreductase family protein	N/A
3,09	Afu3g14750	Fungal specific transcription factor; putative	N/A
2,87	Afu5g02260	ABC multidrug transporter; putative	N/A
2,71	Afu6g08140	Cytochrome P450 monooxygenase; putative	N/A
2,52	Afu4g01440	Glutathione S-transferase family protein	N/A
2,49	Afu3g00560	Putative uncharacterized protein, Has domain(s) with predicted catalytic activity, molybdenum ion binding, pyridoxal phosphate binding activity (	N/A
2,37	Afu1g06480	Putative uncharacterized protein, adhesin	N/A
2,37	Afu5g10060	Cytochrome b5 reductase; putative	N/A
2,34	Afu6g02220	MFS toxin efflux pump; putative	N/A
2,25	Afu5g10050	Cytochrome P450 monooxygenase; putative	N/A
2,21	Afu7g00150	FAD-dependent monooxygenase; putative	nscC
2,21	Afu7g01720	3-hydroxymethyl-3-methylglutaryl-Coenzyme A lyase	N/A
2,19	Afu4g14705	Putative uncharacterized protein	N/A
2,18	Afu3g14630	Extracellular dioxygenase; putative, Has domain(s) with predicted catalytic activity, ferric iron binding, iron ion binding, oxidoreductase activity	N/A
2,14	Afu1g17220	Probable endopolygalacturonase AFUA_1G17220	N/A
2,12	Afu3g08950	Putative uncharacterized protein	N/A
2,09	Afu5g10040	C6 transcription factor; putative	N/A
2,08	Afu2g10240	NAD binding Rossmann fold oxidoreductase; putative	N/A
2,08	Afu6g03380	Putative uncharacterized protein	N/A
2,07	Afu3g00960	Putative uncharacterized protein	N/A
2,04	Afu6g14140	RTA1 domain protein; putative	N/A
1,99	Afu5g10070	3-hydroxyacyl-CoA dehydrogenase; putative	N/A
1,98	Afu7g00470	Putative uncharacterized protein, FungiDB: possible transcription factor	N/A

Table 2. *Cont.*

		Down-Regulated	
-1,09	Afu1g13550	Putative uncharacterized protein	N/A
-1,09	Afu3g14950	Multicopper oxidase; putative	N/A
-1,11	Afu4g08420	Putative uncharacterized protein	N/A
-1,11	Afu3g08990	Cell surface protein; putative	cspA
-1,12	Afu6g10660	ATP citrate lyase subunit (Acl); putatibe	aclA
-1,12	Afu8g01910	Histidine acid phosphatase; putative	N/A
-1,13	Afu3g07560	Enoyl-CoA hydratase/isomerase family protein	N/A
-1,14	Afu6g11310	Bifunctional pyrimidine biosynthesis protein (PyrABCN); putative	N/A
-1,15	Afu4g00860	Dehydrin-like protein	dprA
-1,16	Afu5g02500	Putative uncharacterized protein	N/A
-1,19	Afu4g03240	Cell wall serine-threonine-rich galactomannoprotein Mp1	mp1
-1,19	Afu3g07710	Nucleolin protein Nsr1; putative	N/A
-1,19	Afu3g03940	2;3-diketo-5-methylthio-1-phosphopentane phosphatase; putative	N/A
-1,25	Afu5g00790	ABC multidrug transporter; putative	N/A
-1,25	Afu7g06580	FAD/FMN-containing isoamyl alcohol oxidase MreA-like; putative	N/A
-1,31	Afu6g00510	NADP-dependent alcohol dehydrogenase	N/A
-1,32	Afu7g01730	Phosphatidylserine decarboxylase family protein	N/A
-1,32	Afu7g06350	Sodium/phosphate symporter; putative	phoE
-1,33	Afu2g00625	Putative uncharacterized protein	N/A
-1,34	Afu4g08400	Putative uncharacterized protein	N/A
-1,35	Afu1g04300	Putative uncharacterized protein	N/A
-1,35	Afu5g01630	MFS multidrug transporter; putative	N/A
-1,40	Afu5g14740	Fucose-specific lectin FleA	fleA
-1,42	Afu8g07160	Putative uncharacterized protein	N/A
-2,15	Afu8g05810	DUF1295 domain protein	N/A

2.3. Sterol Metabolism and Regulation

The inhibition of Cyp51A by azole compounds resulted in changes in gene expression in the ergosterol biosynthesis pathway, as shown in previous research [30,33]. In this study, we also assessed the expression levels of genes involved in the ergosterol biosynthesis pathway. When we compared our DMSO-treated isolates of V162(R) and V181(R^{PAN}) with the susceptible DMSO-treated isolate V147(S), the *cyp51A* expression was log2FC 3.12 higher in isolate V162(R) and log2FC 3.46 higher in V181(R^{PAN}), in accordance with the literature as expected [30,37] (Table S6).

Several genes involved in the biosynthesis of mevalonate, an essential precursor substrate for ergosterol synthesis, are up-regulated in V147(S)-ICZ and V162(R)-ISA (Figure 2). *erg13A* and *hmg2* were found up-regulated in both strains, with the mevalonate kinase *erg12* additionally up-regulated log2FC 1.35 in V147(S)-ICZ. Analysis of ergosterol biosynthesis genes showed up-regulation of the majority of *erg* genes in V147(S)-ICZ, V147(S)-ISA, and V162(R)-ISA (Figure 2). No DEGs were found in V181(R^{PAN}) after treatment with ICZ or ISA.

The *cyp51a* eburicol 14α-demethylase and its paralog *cyp51b* were only found to be up-regulated in susceptible isolate V147(S) after both treatments. The *erg6* gene, functioning before Cyp51A in this pathway, was found to be the most up-regulated in both treatments of V147(S) and V162(R). Erg6p was identified as a protein that localizes at lipid droplets in *S. cerevisiae* [37], and a homolog of this protein in *A. nidulans* was found to localize at lipid droplets as well [38,39]. *A. fumigatus* Erg6 shares 88.86% amino acid sequence identity with the *A. nidulans* Erg6 enzyme (NCBI BLASTp). The sterol delta 5;6-desaturase *erg3* (Afu2g00320) was found to be up-regulated in all isolates but the pan-azole-resistant V181(R^{PAN}) isolate.

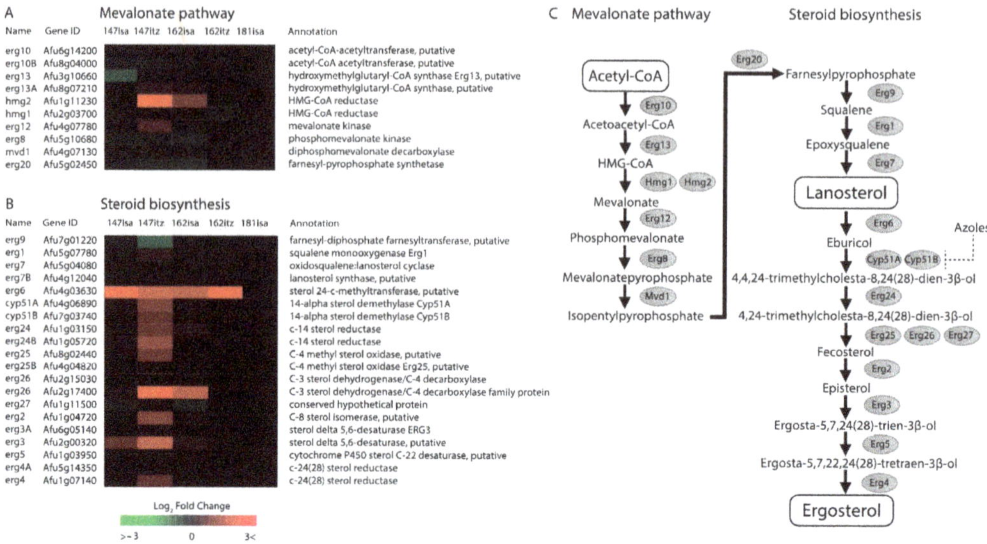

Figure 2. Ergosterol biosynthesis. The colors represent the Log2FC of each time point relative to reference time point t = 0. (**A**). Heatmap of all enzymes with a verified or predicted role in the mevalonate biosynthesis. (**B**). Heatmap of all enzymes with a verified or predicted role in the ergosterol biosynthesis. The *erg6* gene shows strong up-regulation through all isolates and treatments except V180(R^{PAN})-ISA. (**C**). Mevalonate and ergosterol biosynthesis pathways in *A. fumigatus*.

Additional to changes in sterol metabolism, several genes involved in sterol transport and regulation were found to be differentially expressed. In isolates V147(S)-ISA, V147(S)-ICZ, and V162(R)-ISA, we found moderate up-regulation of *oshE* (Afu3g05520), a protein involved in sterol transport, which localizes at contact sites between the ER and the plasma membrane (Table S1) [39]. In isolate V181(R^{PAN})-ISA only, we found up-regulation of putative O-acyltransferase gene Afu2g08380 with log2FC 1.05. This gene is the most homologous to the *S. cerevisiae* ARE2 gene, which is an Acyl-CoA:sterol acyltransferase located in the ER, involved in the esterification of sterols, which prepares them for storage in lipid droplets [40]. Lastly, we found up-regulation of Afu7g01960 in isolates V147(S)-ISA, V147(S)-ICZ, and V162(R)-ISA by log2FC 1.01, log2FC 1.36, and log2FC 1.34, respectively (Table S1). This gene is a putative transcription factor that promotes sterol synthesis and shows 36% amino acid sequence identity with the 'sterol uptake control protein 2' UPC2 in *S. cerevisiae* [41].

2.4. Siderophore Biosynthesis Pathway Is Down-Regulated

The siderophore biosynthesis pathway is involved in extracellular iron uptake in low-iron environments and is regulated by key transcription factors SreA and HapX, which are interconnected in a negative feedback loop (Figure 3A) [40–43]. This pathway shares mevalonate as an important substrate with the ergosterol biosynthesis pathway [44]. Siderophore transporter MirB and the majority of the *sid* genes were down-regulated during both treatments in V147(S) and V162(R) (Figure 3B). Additionally, three genes involved in reductive iron assimilation, *ftrA*, *freB*, and *fetC*, were down-regulated when V147(S) and V162(R) were treated with ISA but not with ICZ. Iron-regulatory transcription factor SreA was found up-regulated by Log2FC 1.67 in isolate V147(S)-ICZ and was only slightly up-regulated in all other isolates. The main transcriptional activator of the siderophore system HapX is found down-regulated during all treatments (Figure 3B).

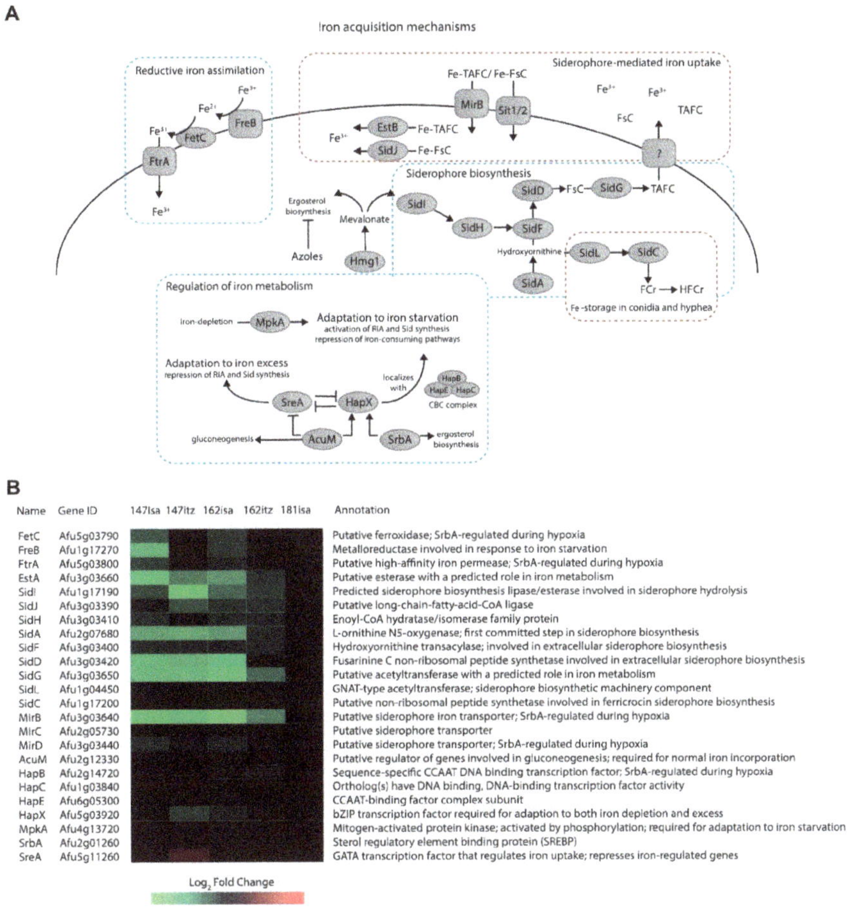

Figure 3. Iron acquisition and siderophore biosynthesis. (**A**). Overview of iron acquisition pathways in *A. fumigatus*. (**B**). Heatmap of genes involved in iron homeostasis.

2.5. Membrane Transporter Regulation

The *A. fumigatus* genome contains various transmembrane transporters that are involved in resistance against antifungal compounds [27]. Therefore, we investigated all verified and putative multidrug-efflux transporters identified in *A. fumigatus* and performed hierarchical clustering on these results.

Verified drug-efflux transporters Mdr1, AbcA, and Cdr1B were significantly up-regulated in all isolates treated with ICZ or ISA (Figure 4). Predominantly down-regulated transporters include AbcB, Mdr4, SitT, Mdr3, and AtrF. Surprisingly, the *abcE* gene was found strongly up-regulated by log2FC 4.31 in the pan-azole-resistant isolate V181(R^{PAN}), whereas this gene was found to be only moderately up-regulated in other isolates. Clustered together with Cdr1b and Mdr1 are putative drug transporters Afu6g03320, Afu2g15140, and Afu2g11580, which show up-regulation in the majority of isolates. Putative MFS transporter Afu6g03320 was found up-regulated in V147(S)-ISA, V147(S)-ICZ, and V162(R)-ISA by log2FC 1.68, 1.74, and 2.35, respectively. Its homolog in *A. nidulans* encodes a sugar transporter [45]. Putative ABC transporter Afu5g10510 was found up-regulated in V147(S)-ICZ, V147(S)-ISA, and V162(R)-ISA, and its homolog in *A. nidulans* encodes a transporter in the cichorine biosynthesis cluster. In V181(R^{PAN})-ISA only, we found up-regulation of

Afu5g02260 with log2FC 2.87, which encodes an ABC transporter homologous to AUS1 and PDR11 in *S. cerevisiae*, two paralogs that facilitate sterol uptake from the environment [46].

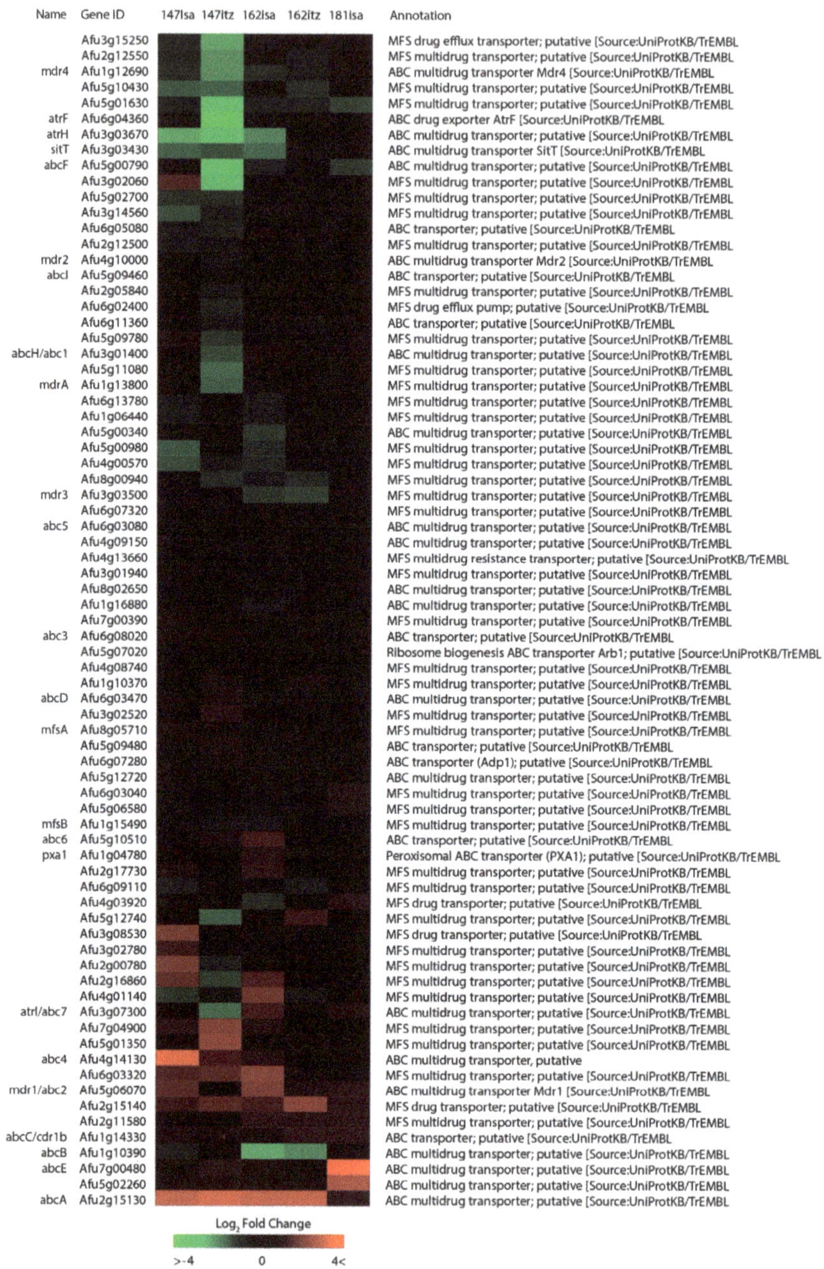

Figure 4. Membrane transporters associated with multidrug efflux. Heatmap showing the hierarchical clustering of the expression patterns of all transporters predicted to play a role in drug efflux in *A. fumigatus*. The colors represent the Log2FC of each time point, relative to reference time point t = 0. Euclidean distances were calculated and complete-linkage hierarchical clustering was performed.

2.6. Secondary Metabolite Cluster Activation

We assessed the 26 secondary metabolite clusters currently identified in A. fumigatus [47]. Four metabolic clusters show enhanced transcription in our dataset after azole treatment (Figure 5). For an overview of all currently identified secondary metabolite gene clusters in A. fumigatus and their corresponding log2FC values during this experiment, see Table S5.

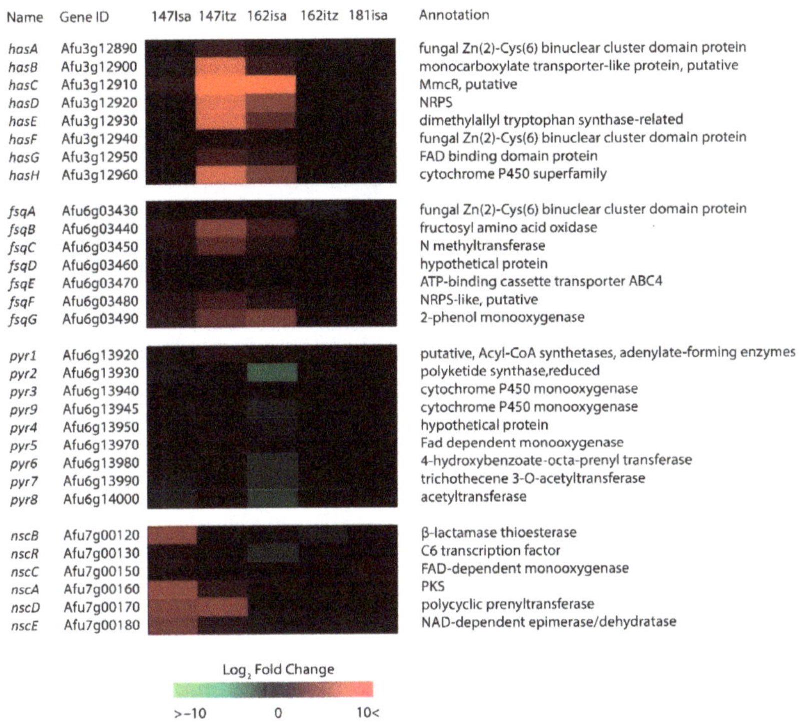

Figure 5. Secondary metabolite production. Heatmap of four secondary metabolite gene clusters that were differently expressed upon addition of ITZ or ISA to the medium.

The *has* cluster consists of eight genes and is involved in hexadehydroastechrome biosynthesis, a molecule with iron-binding properties. An *A. fumigatus* strain with *hasB* overexpression causes a higher mortality rate in a neutropenic murine model [48]. We observed strong up-regulation in the V147(S)-ICZ and V162(R)-ISA isolates. The V147(S)-ISA isolate only showed moderate up-regulation of *hasC*, *hasD*, and *hasH*, whereas no strong regulation was found in all other isolates. The fumisoquin cluster, which consists of *fsqA-G*, was found up-regulated in V147(S)-ICZ, V147(S)-ISA, and V162(R)-ISA. This cluster produces several plant-like isoquinoline alkaloids which can have a variety of functions [49]. Pyripyropene A was initially discovered as a treatment for hypercholesterolemia and atherosclerosis and was shown to have insecticidal properties [50–52]. Interestingly, the pyripyropene A cluster was only found up-regulated in V147(S)-ICZ, whereas it was found down-regulated in V147(S)-ISA and V162(R)-ISA. The neosartoricin/fumicycline A biosynthesis cluster synthesizes an immunosuppressive polyketide, which was shown to have a negative effect on in vitro murine T-cell proliferation [53]. This cluster was found up-regulated in V147(S)-ICZ and V147(S)-ISA.

2.7. Stress Response Pathways

Fungi have several cellular mechanisms to cope with specific environmental stresses and adapt to unfavorable conditions. These mechanisms can be induced by specific conditions or are activated by a wider range of conditions [54]. Expression levels of genes associated with stress response pathways were assessed, such as several cellular stress responses, including high osmolarity adaptation, cell wall integrity maintenance, oxidative stress reduction, and heat shock tolerance [29,55]. We observed several stress-related genes to be affected by azole compounds (Figure 6).

Figure 6. Stress response and signaling pathways. Heatmap of genes with a verified or predicted role in adaptation to several environmental stresses.

The putative glycerol-3-phosphate dehydrogenase GfdA (Afu1g02150) is involved in the osmotic stress response and showed moderate up-regulation in V147(S)-ISA (Log2FC 0.98), V162(R)-ISA (Log2FC 0.70), and V162(R) ICZ (Log2FC 0.73). We found strong downregulation of a putative glyceraldehyde 3-phosphate dehydrogenase (Afu5g01030) in V147(S)-ISA, V147(S)-ICZ, V162(R)-ISA, and V162(R)-ICZ by log2FC −5.60, −5.59, −1.29, and −1.35, respectively. This gene traditionally functions in glycolysis, and its transcription is inhibited in the presence of an oxidant in *S. cerevisiae* [56]. *A. fumigatus* has two orthologous genes, *gpdA* (Afu5g01970) and *gpdB* (Afu8g02560), which are differentially regulated by log2FC −1.05 and log2FC 1.12 in V147(S)-ICZ in our dataset.

The transcription factor AfYap1 is the main regulator of the oxidative stress response in A. fumigatus [57,58] and was moderately up-regulated in isolates V147(S)-ISA (0.51), V147(S)-ICZ (0.39) and V162(R)-ICZ (0.99). Several target genes of AfYap1 with antioxidant properties were up-regulated in isolates V147(S) and V162(R), which will be discussed below. The putative catalase Cat1 (Afu6g03890) was found down-regulated in isolates V147(S)-ICZ (log2FC−1.82), V162(R)-ISA, and in V181(R^{PAN})-ISA (log2FC−1.05). The putative catalase-peroxidase Cat2 (Afu8g01670) was found up-regulated only in isolates V147(S)-ISA and V147(S)-ICZ, with log2FC 0.80 in both treatments. Additionally, the mitochondrial cytochrome c peroxidase gene ccp1 was found up-regulated in V147(S)-ISA and V147(S)-ICZ by log2FC 1.27 and 1.23, respectively. Lastly, the mitochondrial peroxidase gene *prx1* (Afu4g08580) is also regulated by AfYap1 but was down-regulated in V147(S)-ISA, V147(S)-ICZ, and V162(R)-ISA by log2FC −0.86, −1.31, and −2.02, respectively. The putative glutaredoxin Grx1 (Afu1g06100) was found up-regulated in V147(S)-ISA, V162(R)-ISA, and V162(R)-ICZ, by log2FC 1.95, 1.94, and 1.21, respectively. Additionally, we found up-regulation in the thioredoxin allergen Aspf28 (Afu6g10300) in V147(S)-ISA, V147(S)-ICZ, and V162(R)-ISA by log2FC 0.9, 1.76, and 1.37, respectively. We did not find differential regulation of the superoxide dismutase (SOD) genes identified in A. fumigatus, except the putative SOD (Afu1g11640), which was found up-regulated only in isolate V147(S)-ICZ with log2FC 2.13.

2.8. The ER Stress Response

The unfolded protein response (UPR) is activated upon ER stress due to misfolded proteins, induces genes involved in lipid biosynthesis and ER-associated protein degradation (ERAD), and thereby prevents the formation of toxic protein aggregates [59]. Therefore, expression levels of genes associated with the ER stress response were assessed. The ER sensory protein IreA detects unfolded proteins and induces translation of the transcription factor HacA, which activates the expression of target genes with an unfolded protein response element (UPRE) [59,60]. *ireA* and *hacA* were found up-regulated in both treatments of isolates V147(S) and V162(R), with the most up-regulation found in isolate V147(S)-ICZ with log2FC 0.98 and 1.31, respectively (Figure 6). Furthermore, the UPR Hsp70 chaperone BiP/Kar2 that resides in the ER was found up-regulated in V162(R)-ICZ and V181(R^{PAN})-ISA, with log2FC 1.06 and 1.58, respectively. The unfolded protein response (UPR) protein OrmA (Afu4G13270) [60] was found up-regulated in isolate V147(S)-ICZ by log2FC 0.95. This protein is also involved in regulating sphingolipid biosynthesis, an important membrane lipid that associates closely with ergosterol in the membrane [61].

Additionally, the chaperone PdiA, which functions in ER-associated degradation of proteins (ERAD) (Afu2g06150), was found up-regulated by log2FC 1.14 in isolate V147(S)-ICZ. We found several genes present in the ubiquitin-proteasome pathway to be up-regulated. Ubiquitin conjugating enzymes (E2) UbcA (Afu5g07040) and UbcC (Afu5g09200) were found up-regulated in V147(S)-ICZ and V162(R)-ISA. UbcA was found up-regulated in V147(S)-ICZ and V162(R)-ISA by log2FC 1.12 and 1.09, respectively. UbcC was found up-regulated in V147(S)-ICZ by log2FC 1.29. The homolog of an E3 ligase specifically involved in ERAD in *S. cerevisiae*, Doa10 (Afu2g08650) [62], was found up-regulated in isolate V147(S)-ICZ and V162(R)-ISA with log2FC 2.79 and 1.33, respectively. Furthermore, we found up-regulation of two homologs of the cytosolic SCF ubiquitin E3 ligase complex subunit, CulA

(Afu1g12960) and SkpA (Afugg06060), in V147(S) and V162(R) [63]. The fungal-specific F-box protein Fbx15 (Afu3g14150) is part of the SCF complex and is essential during stress adaptation [64]. We found an up-regulation of log2FC 0.91, 1.37, and 1.29 in V147(S)-ICZ, V147(S)-ISA, and V162(R)-ISA, respectively. Furthermore, the transcriptional repressor SsnF (Afu2g11840) was found down-regulated in V147(S)-ICZ by log2FC −1.07, which regulates the derepression of stress response genes and secondary metabolite genes [64].

Another important stress adaptation mechanism comprises the heat shock response, which is induced when environmental temperatures rise but is also activated upon stress conditions that interfere with protein integrity [65]. Heat shock proteins (HSPs) help fold proteins to their final state, aid in the stabilization of unfolded proteins, and prevent the accumulation of protein aggregates [66]. We observed the up-regulation of many essential proteins involved in the heat shock response and protein folding, including the highly conserved cytoplasmic chaperone Hsp70, the Hsp70 chaperone BiP/Kar2, molecular chaperone Hsp90, and Hsp90 co-chaperone Cdc37 (Figure 6). The Hsp70 chaperone BiP/Kar2 that resides in the ER was found up-regulated in V147(S)-ISA, V147(S)-ICZ, V162(R)-ICZ, and V181(R^{PAN})-ISA, as mentioned. We found the most up-regulation of heat-shock-response-related proteins in the pan-azole-resistant isolate V181(R^{PAN}) after ISA treatment. Additionally, we found up-regulation of the mitochondrial Hsp60 chaperone in the V162(R)-ICZ isolate by log2FC 0.80. Lastly, the mitochondrial alpha-ketoglutarate dehydrogenase complex subunit Kgd1 (Afu4g11650) was found up-regulated in V147(S)-ISA, V147(S)-ICZ, V162(R)-ICZ, and V181(R^{PAN})-ISA by log2FC 1.23, 1.61, 1.17, and 1.21, respectively. This protein complex is induced by heat shock conditions, functions to form succinyl-CoA in the TCA cycle, and was found to function as an oxidative stress sensor in mammals [67–69]. Furthermore, it was found to bind specifically to respiratory complex I in mammals [70].

2.9. Gene Ontology (GO) Term Enrichment Analysis

In order to see if any biological processes were significantly altered that were not part of our genomic analysis based on current knowledge of virulence and resistance development, a GO term enrichment analysis was performed; see Table S7. All significant up- and down-regulated DEGs were analyzed separately for the 'Biological process', 'Molecular function', and 'Cellular component' categories.

Many enriched categories in all treated isolates represent broad functional categories or are related to the process of ribosome synthesis and transcription itself, for instance, "GO:0005575 cellular_component, "GO:0002182 cytoplasmic translational elongation", and "GO:0003735 structural constituent of ribosome", which are the most enriched categories of isolate V147(S) for all up- and down-regulated genes after treatment with ICZ or ISA. Enriched GO terms that were found that described more specific cellular processes included "GO:0005739 mitochondrion", "GO:0044550 secondary metabolite biosynthetic process", "GO:0010106 cellular response to iron ion starvation", "GO:0009986 cell surface", "GO:0055085 transmembrane transport", "GO:0051082 unfolded protein binding", and "GO:0034605 cellular response to heat". As no specific trends or overlapping enriched categories were seen between isolates, apart from the categories mentioned above that have been described in the results, we did not assess this data further.

3. Discussion

A. fumigatus has evolved molecular mechanisms to adapt its cellular physiology and overcome stressful environments. Our transcriptome study has provided insights into the transcriptional responses and molecular mechanisms that contribute to the survival of *A. fumigatus* hyphal cells during exposure to azole compounds. The most predominantly differentially regulated processes included sterol metabolism, siderophore-mediated iron acquisition, and secondary metabolite production. Several ABC and MFS transmembrane transporters were up-regulated, which implies increased efflux of cytotoxic compounds and possibly altered nutrient uptake. Furthermore, a variety of genes involved in several evolutionary conserved stress pathways was found up-regulated in varying degrees across tested *A. fumigatus* isolates.

Our exploratory analysis showed that the isolates tested show distinct background expression, as the sample factor appears to have a larger influence on sample distances than the treatment factor, which could be explained by differences in epigenetic regulation between isolates, as this is known to be a large factor in determining transcriptional variation [71]. Additionally, the basal differences in gene expression may also be due to the overexpression of *cyp51A* in the resistant isolates, as this may cause adaptive changes in the expression levels of other genes. Although we observed similarities in gene regulation between treated isolates, many DEGs appear to be specific for one sample and one treatment. This observation is partly explained by genetic variation for transcriptional plasticity, which describes genes that respond differently to changing environments depending on the isolate [72]. Although a general environmental stress response is observed in *S. cerevisiae* [73], this is not clearly observed in *Aspergillus* species to date [74]. In this study, the observed results were not validated by qPCR, as previous experiments by us and other groups have shown the validity of RNA-seq methods for *Aspergillus* spp. ([29,36,75–81]). In addition, the benchmark study conducted by Evereart et al., 2017 (Sci Rep) [82] showed a high concordance (~80%) between RNA-seq and qPCR data for various workflows and mapping tools. Of all non-concordant genes, the majority had a fold change of <2, which led to the observation that only 1.8% of the total gene set (13,045) in this paper showed a significantly different fold change when RNA-seq results were compared to RT-PCR results. In our dataset, we observed the most DEGs in V147(S)-ICZ, V147(S)-ISA, and V162(R)-ISA, suggesting that these isolates suffer the most stress and need to severely alter their metabolism. Overall, we found various similar differentially regulated processes in isolates V147(S) and V162(R) after azole treatments, including increased up-regulation in ergosterol biosynthesis genes as shown in previous research [83]. Isolates V147(S)-ISA, V147(S)-ICZ, and V162(R)-ISA showed up-regulation of the majority of genes involved in ergosterol biosynthesis, whereas itraconazole-resistant isolate V162(R)-ICZ showed only strong up-regulation for the *erg6* gene (Afu4g03630). Our previous study on the immediate transcriptomic response of *A. fumigatus* to itraconazole also showed up-regulation of *erg6* within 30 min of ICZ addition to the medium, which persisted until 240 min after azole addition [36]. The Erg6 enzyme functions directly upstream of Cyp51A in the sterol biosynthesis pathway in *A. fumigatus*, where it methylates lanosterol at C24 to form eburicol [13]. It remains unclear specifically why *erg6* is strongly up-regulated upon azole perturbation, although this was also observed in previous studies [36,83,84]. This enzyme is a potential antifungal target, as Erg6 catalyzes a reaction that is not present in the cholesterol biosynthesis pathway [85].

Previous research showed that repression of *erg11* in a conditional mutant of *Neurospora crassa* results in a similar up-regulation of *erg6* by log2FC 5.1, as well as up-regulation of other *erg* genes in this pathway. The authors show that the transcriptional response of ergosterol biosynthesis genes is not a direct result of azole drug presence or ergosterol depletion but rather the consequence of accumulated sterol intermediates in the cell [83]. However, direct evidence for this theory still has to be provided for *A. fumigatus*. Interestingly, pan-azole-resistant strain V181(R^{PAN}) did not show any differential regulation in ergosterol biosynthesis, which could be explained by the lack of intermediate sterols accumulating in the cell, although further research is required to support this notion. In this isolate, the $TR46^3$ in the promotor region of the *cyp51A* gene is responsible for a strong increase in the Cyp51A enzyme, which possibly ensures a continuous ergosterol biosynthesis despite a high concentration of azoles present in the cell. In addition, the V181(R^{PAN}) isolate showed differential expression of several hundred genes when treated with isavuconazole but not when treated with itraconazole, even though its genotype confers resistance to both azoles. An explanation for this might be that the growth observed in a microtiter plate from which the IC50 concentrations are derived is not an accurate predictor for transcriptome changes taking place in the mycelium, other than the absence of differential regulation in the ergosterol biosynthesis pathway.

The siderophore biosynthesis pathway is involved in iron acquisition when the intracellular iron concentration is low and shares mevalonate as an important substrate with the ergosterol biosynthesis pathway [44]. Adaptation to limited iron availability is essential to maintain fungal virulence [42]. During infections, the innate immune system is able to restrict iron availability to counteract infections, through iron-chelating proteins secreted by macrophages and neutrophils, underlining the importance of these mechanisms in *A. fumigatus* [86].

We observed downregulation of the majority of *sid* genes and siderophore importer MirB in isolates V147(S) and V162(R) during both treatments, implying a metabolic shift for mevalonate as it is preferentially converted to ergosterol upon ergosterol biosynthesis inhibition by azoles. This observation suggests that mevalonate is a key metabolite in susceptible and resistant isolates during azole therapy. Furthermore, these data show that *A. fumigatus* does not suffer from iron depletion under these conditions, as up-regulation of genes involved in siderophore biosynthesis and siderophore-mediated iron uptake would be expected.

The involvement of transmembrane transporters in antifungal resistance has been elucidated over the past decades. *A. fumigatus* has various drug transporters involved in azole resistance, including AbcA (Afu2g15130) and Cdr1B (Afu1g14330) [87]. Our findings included significant up-regulation of these transporters during both ICZ and ISA treatment in all isolates. A recent study on heterologously expressed *A. fumigatus* transporters in a ΔPDR5 *S. cervisiae* isolate showed that expression of *abcA* (Afu2g15130) only resulted in a slight increase in the posaconazole MIC [88], a finding that is in contrast with an earlier study that showed voriconazole sensitization of an *abcA* knockout strain [87]. A recent study created tetracycline-inducible *A. fumigatus* isolates of seven ABC transporters and found increased growth on solid agar supplemented with voriconazole in the strains overexpressing *abc1* (Afu3g01400), *abc3* (Afu6g08020), *abc4* (Afu4g14130), and *abc5* (Afu6g03080) [28]. We found up-regulation of *abc1* in V147(S)-ISA and V162(R)-ISA but downregulation in V147(S)-ICZ, indicating that this gene is not up-regulated by ICZ-induced stress. Esquivel et al. showed that this *abc1* transporter (dubbed *abcH* in their study) had a narrow substrate range and even had a negative impact on the ICZ MIC [88], which could explain the difference in regulation after ICZ or ISA treatment seen in our dataset. In the same study, the expression of *abcF* (Afu5g00790) or *abcC/cdr1B* (Afu1g14330) was found to elevate MICs of all azoles tested, but whereas *abcC* was found up-regulated in all our isolates, *abcF* was found down-regulated in the majority of our isolates. This shows that efflux transporters that can increase the azole MICs of an isolate when artificially up-regulated in the laboratory are not necessarily up-regulated naturally by *A. fumigatus* upon azole addition to the medium.

We also found up-regulation of MFS transporter *mdr1* (Afu5g06070) in all isolates during ICZ and ISA treatment, although a recent study showed that a conditional $Mdr1_{TetOn}$ strain had no influence on voriconazole susceptibility compared to the wildtype [28]. This implies that *mdr1* is activated as the result of a general stress response but is not essential for azole resistance. Interestingly, we found the highest up-regulation of *abcE* (Afu7g00480) with log2FC 4.32 in pan-azole-resistant isolate V181(R^{PAN}) during ISA treatment, whereas this gene was only moderately up-regulated in all other isolates. Previously, *abcE* was found up-regulated in response to voriconazole in the medium [34]. In *A. niger*, a syntenic homolog of this gene is present (An12g03150), which is induced after dithiothreitol (DTT) treatment, a compound that induces the ER stress response [89]. This could indicate that *abcE* is up-regulated due to an activated UPR pathway in isolate V181(R^{PAN}).

Overall, we found differential regulation of many membrane transporters with an unknown function upon azole addition. Recently, it was observed that CDR drug transporters in *C. albicans* were transcriptionally activated upon accumulation of toxic sterol intermediates, an indirect effect resulting from azole treatment [83]. This demonstrates that transporters that are up-regulated upon azole treatment are not necessarily involved in azole efflux directly but are involved in the efflux of accumulated toxic sterol intermediates or are activated by general stress responses.

These results highlight that several transmembrane transporters that were shown to increase the MIC for azole compounds are transcriptionally activated and that general stress response mechanisms can up-regulate transporters that do not provide efflux of azole compounds or directly improve growth in the proximity of azole compounds. Secondary metabolites can provide protection against environmental stresses and microorganisms, and the production of these small bioactive molecules is known to increase virulence in *A. fumigatus* [47,50,90]. Secondary metabolite production upon azole addition in the medium has not been reported to date. We found up-regulation of the fumisoquin biosynthetic cluster (BCG18) in isolates V147(S)-ISA, V147(S)-ICZ, and V162(R)-ISA, indicating that these compounds can be produced in susceptible isolates and resistant isolates upon azole treatment. Overexpression of the transcriptional regulator *fsqA* did not increase virulence in a neutropenic murine model [91]. The *has* cluster was up-regulated in V147(S)-ICZ and V162(R)-ISA, indicating that the cell increases iron-binding hexadehydroastechrome production, perhaps to counter a down-regulated siderophore production, which limits iron acquisition. The has cluster was also found to be up-regulated after iron starvation and H_2O_2-induced oxidative stress [92]. For reasons unknown, up-regulation of the pyripyropene A biosynthetic cluster was only found in V147(S)-ICZ, in contrast to isolates V147(S)-ISA and V162(R)-ISA where this cluster was found to be down-regulated. Up-regulation of neosartoricin/fumicycline A biosynthetic cluster was solely found in isolate V147(S), indicating that the production of this compound after azole addition is specific for this isolate. Interestingly, we found up-regulation of several genes present in the gliotoxin biosynthetic cluster in V147(S)-ISA, V147(S)-ICZ, and V162(R)-ICZ, although this up-regulation could not be rendered as significant in many instances due to a low transcript count (Table S5). Previous research has shown that *A. fumigatus* produces more gliotoxin when grown in a biofilm than when grown planktonic in a liquid medium, which could explain why we observed low quantities of gliotoxin-related gene transcripts [93]. We show that azole treatment of susceptible and resistant isolates can result in the up-regulation of several secondary metabolite-encoding biosynthetic clusters.

Several antifungal compounds have been reported to induce the formation of reactive oxygen species (ROS) in the cell as a secondary mechanism of antifungal activity. Intracellular ROS causes lipid peroxidation, which ultimately threatens the integrity of membranes [94]. Among the azole class antifungals, micafungin was shown to induce ROS species in *C. albicans* biofilms [95], and itraconazole was shown to induce ROS in *A. fumigatus*, although to a lesser extent than terbinafine and amphotericin B [96]. In our dataset, we observed up-regulation of transcription factor Yap1 (Afu6g09930) in isolate V162(R)-ICZ, which translocates from the cytoplasm to the nucleus upon oxidative stress [26]. We also observe up-regulation of catalases CatA (Afu6g03890) in isolate V162(R)-ICZ; Cat2 (Afu8g01670) in isolates V147(S)-ISA and V147(S)-ICZ; uncharacterized glutaredoxin 'GRX1' (Afu1g06100) in the majority of isolates; glutaredoxin *grxD* (Afu2g14960) in isolates V147(S)-ISA, V147(S)-ICZ, and V162(R)-ICZ; the mitochondrial cytochrome C peroxidase (Afu4g09110) in isolates V147(S)-ISA and V147(S)-ICZ; an uncharacterized superoxide dismutase (Afu1g11640); and the putative thioredoxin Aspf28 (Afu6g10300) in V147(S)-ISA, V147(S)-ICZ, and V162(R)-ISA. This indicates that these isolates experience oxidative stress upon azole addition to the medium. This observation corresponds with the hypothesis that ROS are produced by membrane-perturbating compounds that cause the destabilization of membranes and the mitochondrial complex I [96]. However, the up-regulation of these genes was found to be variable between isolates and treatments. In *A. fumigatus*, the SOD genes have been observed to mainly counteract oxidants, which naturally arise in the cell due to metabolism and respiration, rather than detoxify ROS as part of a stress response pathway [97]. Instead, thioredoxins and glutaredoxins could be playing a larger role in detoxifying stress-induced ROS. We found up-regulation of an uncharacterized homolog of

GRX1 (Afu1g06100) and the essential glutaredoxin *grxD* (Afu2g14960); the latter was found to be up-regulated under iron starvation conditions when it functions to activate HapX. The overexpression of this gene was not found to influence other genes under iron-replete conditions [92].

We found differential regulation of genes involved in the ER stress response and the unfolded protein response (UPR), which is activated upon ER stress. Upon ER stress, the expression of constitutively active heat shock proteins is up-regulated, which are essential in facilitating protein folding under normal conditions and protecting the cell by stabilizing unfolded proteins [69].

We found up-regulation of *ireA* (Afu1g01720) and *hacA* (Afu3G04070) in isolates V147(S) and V162(R) during both treatments. Furthermore, we found up-regulation of UPR protein *ormA* (Afu4g13270) in isolate V147(S)-ICZ. It was shown that the overexpression of *ormA* in *A. fumigatus* leads to reduced ICZ susceptibility [61].

The ER-resident HSP70 chaperone BiP/KAR2 is involved in protein folding, protein translocation across the ER membrane, and ER-associated degradation and was found up-regulated in the majority of isolates. In the pathogenic yeast *Cryptococcus neoformans*, a transcriptomics approach uncovered that azole treatment is sensed as ER stress, which activates Ire1 and results in the induction of KAR2 by the UPR pathway [98]. Indeed, induction of KAR2 was found to be essential to dispose of misfolded proteins, which alleviates ER stress [99]. Another key stress response protein Hsp90 (Afu5g04170) was found up-regulated, which is mainly involved in the final maturation of proteins and the assembly of complex macromolecular structures [66]. Furthermore, Hsp90 co-chaperone Cdc37 (Afu4g10010) and Hsp70/Hsp90 co-chaperone Sti1 (Afu7g01860) were found up-regulated, which are involved in the stabilization of kinases and protein aggregates and suppression of proteotoxicity from aggregated proteins, respectively [100]. In *T. rubrum*, *hsp90* deletion was found to have a profound effect on itraconazole tolerance [101]. Furthermore, DSC complex subunit ubiquitin ligase *dscA* (Afu1g12080) present in the Golgi apparatus was also found up-regulated in isolates V147(S) and V162(R) during both treatments. This complex is critical for the activation of SrbA by proteolytic cleavage and thus for the cellular response to azole drugs [102]. Lastly, the highly conserved ERAD-associated E3 ubiquitin ligase Doa10 was found up-regulated in isolates V147(S) and V162(R) during both treatments, which facilitates the degradation of Erg1 after lanosterol accumulation in order to prevent further accumulation of disruptive sterol intermediates in the cell [103].

4. In Summary

It appears that the transcriptome response is in part determined by the genomic background of the isolate. Transcriptional plasticity was shown to be more variable in multi-copy and non-essential genes [72], which might further influence the transcriptional response we observed. This underlines the importance of the characterization of genes as it enables us to better understand physical adaptation through transcriptional regulation. These data show that the pan-azole-resistant isolate with the highest MICs for azoles shows the least amount of DEGs when confronted with azole compounds, and our susceptible isolate V147(S) with the lowest MICs for azoles displayed the highest amount of DEGs, with ICZ-resistant isolate V162(R) being in the middle.

We found that many HSPs were up-regulated in pan-azole-resistant isolate V181(R^{PAN}) harboring a $TR46^3$, originally isolated from the environment [30]. It is possible that hyperactivation of *cyp51A* transcription under azole growth pressure causes increased protein misfolding and ER stress; hence, HSPs are activated to alleviate ER stress [104]. These observations fall in line with the strong up-regulation that was found in this isolate for *abcE*, a transporter that was shown to be up-regulated in *A. niger* after DTT-induced ER stress [89]. Additionally, it could reflect adaptation to the environment that this strain has been isolated in, a compost heap that can reach temperatures of up to 70 °C [30]. Interestingly, this mechanism has been observed before in thermotolerant yeasts, that global transcription changes as a 'preventive response' contributed to growth at elevated temperatures [105].

Interestingly, we found no differential regulation of *cyp51A* or any other genes involved in ergosterol biosynthesis in V181(R^{PAN}) after azole addition to the medium, underlining that there might be no disruption in ergosterol biosynthesis in this strain when confronted with isavuconazole, as compared to the untreated isolate. However, as the base-level expression of hundreds of genes including *cyp51A* in isolate V181(R^{PAN}) are differentially expressed when compared to isolate V147(S) as shown in Table S6, it is plausible that this isolate is already equipped to encounter azole compounds in the environment, which would remove the need for *A. fumigatus* to further alter its transcription of ergosterol biosynthesis genes.

Considering azole-resistant *A. fumigatus* isolates, many factors that determine the resistance phenotype remain unknown. We found diverse transcriptomes after treating three *A. fumigatus* isolates with a distinct genetic background with azole antifungal compounds. Although several cellular processes were subjected to similar regulation, including ergosterol biosynthesis up-regulation and siderophore biosynthesis downregulation, many isolates have a transcriptome unique for that isolate.

Therefore, it is important that the transcriptional response of more clinical and environmental isolates is measured. Additional RNA-seq data could provide information on the diversity of the transcriptional response when *A. fumigatus* is confronted with azole compounds to uncover possible core gene sets that are transcriptionally adjusted in isolates with similar genetic backgrounds or comparable azole susceptibilities.

5. Conclusions

Several cellular mechanisms are involved in the fungal stress response against azole compounds, and determining their contribution to the resistant phenotype is valuable in being able to recognize more targets for antifungal drug development and optimize azole therapy.

Monitoring transcriptome changes could help us understand which changes in gene expression have a strong influence on the phenotype and which transcriptional changes allow for better adaptation and survival of the fungus in the presence of azoles.

6. Material and Methods

6.1. Isolates, Media, and Culture Conditions

Three isolates with variable susceptibilities for itraconazole (Sigma-Aldrich, Zwijndrecht, The Netherlands) and isavuconazole (Sigma-Aldrich, Zwijndrecht, The Netherlands) were selected, with isolates V147(S) and V162(R) of clinical origin and pan-azole-resistant isolate V181(R^{PAN}) of environmental origin. All isolates are listed in Table 1. Isolates were cultured on either Sabouraud Dextrose Agar (SDA) (Oxoid, Landsmeer, The Netherlands) (1% peptone, 4% glucose, 1.5% agar, pH 5.6) or Aspergillus Minimal Medium (AMM) as described in the supplemental material of [106] containing per liter: 10 g glucose, 5.95 g $NaNO_3$, 0.522 g KCl, 1.5 g KH_2PO_4, 50 mg $MgSO_4{\cdot}7H_2O$, and 1 mL trace elements. Trace elements contained (per 200 mL): 10 g EDTA, 4.4 g $ZnSO_4{\cdot}7H_2O$, 1.01 g $MnCl_2{\cdot}4H2O$, 0.315 g $CuSO_4{\cdot}5H_2O$, 0.22 g $(NH4)6Mo_7O_{24}{\cdot}4H_2O$, 1.0 g Fe(II)- $SO_4{\cdot}7H_2O$, and 2.2 g H_3BO_3 [107]. All compounds were produced by Merck (Darmstadt, Germany). Isolates were stored in 20% glycerol at −80 °C and subcultured on SDA at 37 °C for 4–5 days. Conidia were harvested with a wet cotton swab and resuspended in Milli-Q containing 0.1% Tween 20. Approximately 2.6×10^5 CFU/mL of conidia were used to inoculate 5 mL of AMM in 15 mL tubes, containing IC_{50} concentrations of ICZ or ISA or the same volume of DMSO as a control. Conidia were inoculated in duplo for each condition. After 48 h of growth at 37 °C while rotating, mycelial bulbs were harvested through Miracloth (Merck) and directly frozen in liquid nitrogen.

6.2. Dose–Response Curves

For determining the ICZ and ISA concentrations where growth is estimated to be 50% (IC50), dose–response curves were made by growing each strain (n = 5) in a 96-well plate containing AMM with itraconazole concentrations ranging from 0.016 µg/mL to 16 µg/mL.

Plates were kept at 37 °C in a microtiter plate reader (Anthos Labtec Instruments GmbH, Salzburg, Austria) for 48 h, and optical density (OD) was measured every hour at 405 nm. After subtraction of the ODs of the empty wells, the percentage of growth from the ODs of the inoculated wells for each well was correlated with the relative OD estimated by using the following equation: (OD405 of wells that contained itraconazole/OD405 of the itraconazole-free well) × 100. The relationship was determined by anon-linear regression analysis and the Hill equation with a variable slope fitted to the data. Goodness of fit was checked by the R^2 values. All analyses were performed by using GraphPad Prism, version 6.0, for Windows (GraphPad Software, San Diego, CA, USA). For isolate V181(R^{PAN}), a treatment concentration of 16 µg/mL ISA was used due to the 0 DEGs that were found after 8 µg/mL ICZ.

6.3. RNA-Sequencing

After harvesting, mycelia were lysed by bead-bashing using a MagNa Lyser instrument three times for 30 s at 7000 rpm (Roche, Basel, Switzerland). RNA was extracted using TRIzol reagent (Invitrogen, Breda, the Netherlands). Pellets were resuspended in 100 µL of DEPC-treated water. RNA integrity was verified through agarose gel electrophoresis. Total RNA was stained with Midori Green (Nippon Genetics, Dueren, Germany) and visualized with UV light. The purity of all samples (A260/A280 $\geq$ 1.5 and A260/A230 $\geq$ 2.0) was determined using a NanoDrop 1000 (Thermo Scientific, MA, USA). Pooled cDNA libraries were constructed with the TruSeq RNA Sample Prep Kit v2 (Illumina, San Diego, CA, USA) according to manufacturer instructions. The quality of the libraries was verified with an Agilent 4200 Tapestation System (Agilent Technologies). For our ICZ-treated isolates, paired-end reads of 2 × 75 bp were generated in High Output mode on a NextSeq 500 sequencer (Illumina, San Diego, CA, USA) at the Genome Diagnostics and the Core Genome Analysis Laboratory of Radboudumc. For our ISA-treated isolates, paired-end reads of 2 × 100 bp were generated on a HiSeq 4000 (Illumina, San Diego, CA, USA) at BGI Genomics (Copenhagen, Denmark).

6.4. Identification of Differentially Expressed Genes

Reads were aligned with STAR v2.5.3A [108] against the reference genome sequence of *A. fumigatus* Af293 Ensemble CADRE 30 downloaded from Ensembl Fungi. A batch correction was additionally performed by using DESeq2 in programming language R by Bioconductor [109]. Differentially expressed genes (DEGs) were identified by DESeq2 (BaseMean $\geq$ 36, p-value < 0.05, adjusted p-value < 0.1, log2FC > 0.7 or <−0.7), and results were corrected for multiple testing by the Benjamini–Hochberg principle. A batch correction was additionally performed by DESeq2 to correct for sequencing artifacts created by the difference in batches. Treatment with ISA and ICZ was performed in two separate batches; hence, there are four replicates total of the untreated (DMSO) condition per isolate. The untreated isolates per strain were grouped together for the DESeq2 analysis to increase statistical power. Functional categories for filamentous fungi, including *A. fumigatus*, remain poorly defined. Therefore, a more directional approach was used by assessing pathways associated with stress and virulence based on current knowledge.

GO term analysis was performed in FungiFun V2.2.8 [110]. Fisher's exact test was performed with a Benjamini–Hochberg procedure for multiple-testing error correction for enriched categories in the 'Biological process', 'Cellular component', and 'Molecular function' ontologies. Statistical analysis, heatmaps, and volcano plots were generated in Rstudio version 1.2.1335 for Windows 10. For generating heatmaps, gplots v3.0.1 was used. For generating volcano plots, ggplot2 V2.2.1 was used. Exported images were assembled with Adobe Illustrator CC 2015. Venn diagrams were created with the Venn tool by (http://bioinformatics.psb.ugent.be/webtools/Venn, accessed on 22 March 2022). Graphs of dose–response curve analyses were generated by using GraphPad Prism, version 6.0, for Windows (GraphPad Software, San Diego, CA, USA).

6.5. Microsatellite Genotyping

Short Tandem Repeat (STR) analysis was performed on the isolates used in this study.

Multiplex PCR was performed for *A. fumigatus* STR loci; three trinucleotide (STRAf3A-C) and three tetranucleotide loci (STRAf4A-C) were amplified using corresponding primers containing the dye carboxyfluorescein (FAM), hexachlorofluorescein (HEX), or tetrachlorofluorescein (TET) at the 5′ end [111]. Cleaned PCR products were analyzed together with a GeneScan500 LIZ size standard (Applied Biosystems, Foster City, CA, USA), and fluorescence was detected on a Genetic Analyzer (Applied Biosystems, Foster City, CA, USA). The number of repeats was determined using the Peak Scanner Software v1.0 (Applied Biosystems, Foster City, CA, USA).

Supplementary Materials: The following supporting information can be downloaded at: https://www.mdpi.com/article/10.3390/jof9080807/s1, Figure S1: Colony morphology of *A. fumigatus* isolates used in this study, Reference isolate CEA10 was added for comparison, Isolates were grown for 4d at 37 °C on AMM; Table S1: Complete list of DEseq2 results; Table S2: List of the 50 DEGs with the highest or lowest log2FC for each isolate; Table S3: List of DEGs after DMSO treatment; Table S4: List of DEGs found after Venn analysis; Table S5: All 26 secondary metabolite clusters in A. fumigatus and corresponding log2FC values; Table S6—Analysis of base expression levels; Table S7: Gene ontology enrichment analysis.

Author Contributions: Conceptualization, M.W.J.H., J.Z., P.E.V. and W.J.G.M.; methodology, M.W.J.H., J.Z., J.P.M.C. and W.J.G.M.; experiments and data analysis, M.W.J.H., H.S. and J.P.M.C.; writing—original draft preparation, M.W.J.H. and W.J.G.M.; writing—review and editing, All authors. All authors have read and agreed to the published version of the manuscript.

Funding: This research did not receive any specific grant from funding agencies in the public, commercial, or not-for-profit sectors.

Institutional Review Board Statement: Not applicable.

Informed Consent Statement: Not applicable.

Data Availability Statement: The sequences reported in this paper have been deposited in the NCBI Sequence Read Archive (SRA) database in a BioProject under accession number PRJNA894836.

Acknowledgments: We would like to thank the Department of Genome Diagnostics and the Core Genome Analysis Laboratory of Radboudumc for performing the NextSeq500 sequencing. We would also like to thank Hein van der Lee and Marlou Tehupeiory-Kooreman for their excellent technical assistance.

Conflicts of Interest: The authors declare that they have no competing interests.

References

1. Buil, J.B.; Hare, R.K.; Zwaan, B.J.; Arendrup, M.C.; Melchers, W.J.G.; Verweij, P.E. The fading boundaries between patient and environmental routes of triazole resistance selection in *Aspergillus fumigatus*. *PLoS Pathog.* **2019**, *15*, e1007858. [CrossRef]
2. Chowdhary, A.; Sharma, C.; Meis, J.F. Azole-resistant aspergillosis: Epidemiology, molecular mechanisms, and treatment. *J. Infect. Dis.* **2017**, *216* (Suppl. S3), S436–S444. [CrossRef]
3. Latgé, J.-P.; Chamilos, G. *Aspergillus fumigatus* and Aspergillosis in 2019. *Clin. Microbiol. Rev.* **2020**, *33*, e00140-18. [CrossRef]
4. De Castro, P.A.; Colabardini, A.C.; Moraes, M.; Horta, M.A.C.; Knowles, S.L.; Raja, H.A.; Oberlies, N.H.; Koyama, Y.; Ogawa, M.; Gomi, K.; et al. Regulation of gliotoxin biosynthesis and protection in *Aspergillus* species. *PLoS Gen.* **2022**, *18*, e1009965. [CrossRef]
5. Kolwijck, E.; van de Veerdonk, F.L. The potential impact of the pulmonary microbiome on immunopathogenesis of *Aspergillus*-related lung disease. *Eur. J. Immunol.* **2014**, *44*, 3156–3165. [CrossRef]
6. Cai, L.; Gao, P.; Wang, Z.; Dai, C.; Ning, Y.; Ilkit, M.; Xue, X.; Xiao, J.; Chen, C. Lung and gut microbiomes in pulmonary aspergillosis: Exploring adjunctive therapies to combat the disease. *Front. Immunol.* **2022**, *13*, 988708. [CrossRef]
7. Verweij, P.E.; Brüggemann, R.J.M.; Wauters, J.; Rijnders, B.J.A.; Chiller, T.; van de Veerdonk, F.L. Influenza Coinfection: Be (a)ware of Invasive Aspergillosis. *Clin. Infect. Dis.* **2020**, *70*, 347–349. [CrossRef]
8. Koehler, P.; Bassetti, M.; Chakrabarti, A.; Chen, S.C.A.; Colombo, A.L.; Hoenigl, M.; Klimko, N. Defining and managing COVID-19-associated pulmonary aspergillosis: The 2020 ECMM/ISHAM consensus criteria for research and clinical guidance. *Lancet Infect. Dis.* **2021**, *21*, e149–e162. [CrossRef]

9. Kwon-Chung, K.J.; Sugui, J.A. *Aspergillus fumigatus*—What Makes the Species a Ubiquitous Human Fungal Pathogen? *PLoS Pathog.* **2013**, *9*, e1003743. [CrossRef]
10. Hokken, M.W.J.; Zwaan, B.J.; Melchers, W.J.G.; Verweij, P.E. Facilitators of adaptation and antifungal resistance mechanisms in clinically relevant fungi. *Fungal Genet. Biol.* **2019**, *132*, 103254. [CrossRef]
11. Buil, J.B.; Snelders, E.; Denardi, L.B.; Melchers, W.J.G.; Verweij, P.E. Trends in Azole Resistance in *Aspergillus fumigatus*, the Netherlands, 1994–2016. *Emerg. Infect. Dis.* **2019**, *25*, 176–178. [CrossRef]
12. Zhang, J.; Van Den Heuvel, J.; Debets, A.J.M.; Verweij, P.E.; Melchers, W.J.G.; Zwaan, B.J.; Schoustra, S.E. Evolution of cross-resistance to medical triazoles in *Aspergillus fumigatus* through selection pressure of environmental fungicides. *Proc. R. Soc. B Biol. Sci.* **2017**, *284*, 20170635. [CrossRef]
13. Schoustra, S.E.; Debets, A.J.M.; Rijs, A.J.M.M.; Zhang, J.; Snelders, E.; Leendertse, P.C.; Melchers, W.J.G.; Rietveld, A.G.; Zwaan, B.J.; Verweij, P.E. Environmental hotspots for azole resistance selection of aspergillus fumigatus, the netherlands. *Emerg. Infect. Dis.* **2019**, *25*, 1347–1353. [CrossRef]
14. Snelders, E.; Huis In't Veld, R.A.G.; Rijs, A.J.M.M.; Kema, G.H.J.; Melchers, W.J.G.; Verweij, P.E. Possible environmental origin of resistance of *Aspergillus fumigatus* to medical triazoles. *Appl. Environ. Microbiol.* **2009**, *75*, 4053–4057. [CrossRef]
15. Alcazar-Fuoli, L.; Mellado, E.; Garcia-Effron, G.; Lopez, J.F.; Grimalt, J.O.; Cuenca-Estrella, J.M.; Rodriguez-Tudela, J.L. Ergosterol biosynthesis pathway in *Aspergillus fumigatus*. *Steroids* **2008**, *73*, 339–347. [CrossRef]
16. Hartmann, T.; Sasse, C.; Schedler, A.; Hasenberg, M.; Gunzer, M.; Krappmann, S. Shaping the fungal adaptome—Stress responses of *Aspergillus fumigatus*. *Int. J. Med. Microbiol.* **2011**, *301*, 408–416. [CrossRef]
17. De Vries, R.P.; Riley, R.; Wiebenga, A.; Aguilar-Osorio, G.; Amillis, S.; Uchima, C.A.; Anderluh, G.; Asadollahi, M.; Askin, M.; Barry, K.; et al. Comparative genomics reveals high biological diversity and specific adaptations in the industrially and medically important fungal genus *Aspergillus*. *Genome Biol.* **2017**, *18*, 28. [CrossRef]
18. Verweij, P.E.; Zhang, J.; Debets, A.J.M.; Meis, J.F.; van de Veerdonk, F.L.; Schoustra, S.E.; Zwaan, B.J.; Melchers, W.J.G. In-host adaptation and acquired triazole resistance in *Aspergillus fumigatus*: A dilemma for clinical management. *Lancet Infect. Dis.* **2016**, *16*, e251–e260. [CrossRef]
19. Camps, S.M.T.; Dutilh, B.E.; Arendrup, M.C.; Rijs, A.J.M.M.; Snelders, E.; Huynen, M.A.; Verweij, P.E.; Melchers, W.J.G. Discovery of a hapE Mutation That Causes Azole Resistance in *Aspergillus fumigatus* through Whole Genome Sequencing and Sexual Crossing. *PLoS ONE* **2012**, *7*, e50034. [CrossRef]
20. Hortschansky, P.; Misslinger, M.; Mörl, J.; Gsaller, F.; Bromley, M.J.; Brakhage, A.A.; Groll, M.; Haas, H.; Huber, E.M. Structural basis of HapEP88L-linked antifungal triazole resistance in *Aspergillus fumigatus*. *Life Sci. Alliance* **2020**, *3*, e202000729. [CrossRef]
21. Sewell, T.R.; Zhu, J.; Rhodes, J.; Hagen, F.; Meis, J.F.; Fisher, M.C.; Jombart, T. Nonrandom distribution of azole resistance across the global population of aspergillus fumigatus. *mBio* **2019**, *10*, e00392-19. [CrossRef]
22. Furukawa, T.; van Rhijn, N.; Fraczek, M.; Gsaller, F.; Davies, E.; Carr, P.; Gago, S.; Fortune-Grant, R.; Rahman, S.; Gilsenan, J.M.; et al. The negative cofactor 2 complex is a key regulator of drug resistance in *Aspergillus fumigatus*. *Nat. Commun.* **2020**, *11*, 427. [CrossRef]
23. Paul, S.; Stamnes, M.; Thomas, G.H.; Liu, H.; Hagiwara, D.; Gomi, K.; Filler, S.G.; Moye-Rowley, W.S. AtrR Is an Essential Determinant of Azole Resistance in *Aspergillus fumigatus*. *mBio* **2019**, *10*, e02563-18. [CrossRef]
24. Buil, J.B.; Hagen, F.; Chowdhary, A.; Verweij, P.E.; Meis, J.F. Itraconazole, voriconazole, and posaconazole CLSI MIC distributions for wild-type and Azole-resistant *Aspergillus fumigatus* isolates. *J. Fungi* **2018**, *4*, 103. [CrossRef]
25. Wei, X.; Chen, P.; Gao, R.; Li, Y.; Zhang, A.; Liu, F.; Lu, L. Screening and Characterization of a Non-cyp51A Mutation in an *Aspergillus fumigatus* cox10 Strain Conferring Azole Resistance. *Antimicrob. Agents Chemother.* **2017**, *61*, e02101-16. [CrossRef]
26. Moye-Rowley, W.S. Multiple mechanisms contribute to the development of clinically significant azole resistance in *Aspergillus fumigatus*. *Front. Microbiol.* **2015**, *6*, 70. [CrossRef]
27. Fraczek, M.G.; Bromley, M.; Buied, A.; Moore, C.B.; Rajendran, R.; Rautemaa, R.; Ramage, G.; Denning, D.W.; Bowyer, P. The cdr1B efflux transporter is associated with non-cyp51a-mediated itraconazole resistance in *Aspergillus fumigatus*. *J. Antimicrob. Chemother.* **2013**, *68*, 1486–1496. [CrossRef]
28. Sturm, L.; Geißel, B.; Martin, R.; Wagener, J. Differentially Regulated Transcription Factors and ABC Transporters in a Mitochondrial Dynamics Mutant Can Alter Azole Susceptibility of *Aspergillus fumigatus*. *Front. Microbiol.* **2020**, *11*, 1017. [CrossRef]
29. Takahashi, H.; Kusuya, Y.; Hagiwara, D.; Takahashi-Nakaguchi, A.; Sakai, K.; Gonoi, T. Global gene expression reveals stress-responsive genes in *Aspergillus fumigatus* mycelia. *BMC Genomics* **2017**, *18*, 942. [CrossRef]
30. Zhang, J.; Snelders, E.; Zwaan, B.J.; Schoustra, S.E.; Meis, J.F.; Van Dijk, K.; Hagen, F.; Van Der Beek, M.T.; Kampinga, G.A.; Zoll, J.; et al. A novel environmental azole resistance mutation in *Aspergillus fumigatus* and a possible role of sexual reproduction in its emergence. *MBio* **2017**, *8*, e00791-17. [CrossRef]
31. Nierman, W.C.; Pain, A.; Anderson, M.J.; Wortman, J.R.; Kim, H.S.; Arroyo, J.; Berriman, M.; Abe, K.; Archer, D.B.; Bermejo, C.; et al. Genomic sequence of the pathogenic and allergenic filamentous fungus *Aspergillus fumigatus*. *Nature* **2005**, *438*, 1151–1156. [CrossRef]
32. Hong, L.; Lee, M.J.; Solis, N.V.; Phan, Q.T.; Swidergall, M.; Ralph, B.; Ibrahim, A.S.; Sheppard, D.C.; Filler, S.G. *Aspergillus fumigatus* CalA Binds to Integrin $\alpha5\beta1$ and Mediates Host Cell Invasion. *Nat. Microbiol.* **2017**, *2*, 16211. [CrossRef]

33. Feng, X.; Krishnan, K.; Richie, D.L.; Aimanianda, V.; Hartl, L.; Grahl, N.; Powers-Fletcher, M.V.; Zhang, M.; Fuller, K.K.; Nierman, W.C.; et al. Haca-independent functions of the ER stress sensor irea synergize with the canonical UPR to influence virulence traits in *Aspergillus fumigatus*. *PLoS Pathog.* **2011**, *7*, e1002330. [CrossRef]
34. Da Silva Ferreira, M.E.; Malavazi, I.; Savoldi, M.; Brakhage, A.A.; Goldman, M.H.S.; Kim, H.S.; Nierman, W.C.; Goldman, G.H. Transcriptome analysis of *Aspergillus fumigatus* exposed to voriconazole. *Curr. Genet.* **2006**, *50*, 32–44. [CrossRef]
35. Whaley, S.G.; Tsao, S.; Weber, S.; Zhang, Q.; Barker, K.S.; Raymond, M.; Rogers, P.D. The RTA3 gene, encoding a putative lipid translocase, influences the susceptibility of Candida albicans to fluconazole. *Antimicrob. Agents Chemother.* **2016**, *60*, 6060–6066. [CrossRef]
36. Hokken, M.W.J.; Zoll, J.; Coolen, J.P.M.; Zwaan, B.J.; Verweij, P.E.; Melchers, W.J.G. Phenotypic plasticity and the evolution of azole resistance in *Aspergillus fumigatus*; An expression profile of clinical isolates upon exposure to itraconazole. *BMC Genomics* **2019**, *20*, 28. [CrossRef]
37. Leber, R.; Zinser, E.; Paltauf, F.; Daum, G.; Zellnig, G. Characterization of lipid particles of the yeast, Saccharomyces cerevisiae. *Yeast* **1994**, *10*, 1421–1428. [CrossRef]
38. Salogiannis, J.; Christensen, J.R.; Aguilar-Maldonado, A.; Shukla, N.; Reck-Peterson, S.L. Regulation of peroxisome and lipid droplet hitchhiking by PxdA and the DipA phosphatase. *bioRxiv* **2020**. [CrossRef]
39. Bühler, N.; Hagiwara, D.; Takeshita, N. Functional analysis of sterol transporter orthologues in the filamentous fungus Aspergillus nidulans. *Eukaryot. Cell* **2015**, *14*, 908–921. [CrossRef]
40. Renne, M.F.; Klug, Y.A.; Carvalho, P. Lipid droplet biogenesis: A mystery "unmixing"? *Semin. Cell Dev. Biol.* **2020**, *108*, 14–23. [CrossRef]
41. Ballard, E.; Weber, J.; Melchers, W.J.G.; Tammireddy, S.; Whitfield, P.D.; Brakhage, A.A.; Brown, A.J.P.; Verweij, P.E.; Warris, A. Recreation of in-host acquired single nucleotide polymorphisms by CRISPR-Cas9 reveals an uncharacterised gene playing a role in *Aspergillus fumigatus* azole resistance via a non-cyp51A mediated resistance mechanism. *Fungal Genet. Biol.* **2019**, *130*, 98–106. [CrossRef] [PubMed]
42. Haas, H. Iron—A key nexus in the virulence of *Aspergillus fumigatus*. *Front. Microbiol.* **2012**, *3*, 28. [CrossRef] [PubMed]
43. Haas, H. Fungal siderophore metabolism with a focus on *Aspergillus fumigatus*. *Nat. Prod. Rep.* **2014**, *31*, 1266–1276. [CrossRef] [PubMed]
44. Yasmin, S.; Alcazar-Fuoli, L.; Grundlinger, M.; Puempel, T.; Cairns, T.; Blatzer, M.; Lopez, J.F.; Grimalt, J.O.; Bignell, E.; Haas, H. Mevalonate governs interdependency of ergosterol and siderophore biosyntheses in the fungal pathogen *Aspergillus fumigatus*. *Proc. Natl. Acad. Sci. USA* **2012**, *109*, E497–E504. [CrossRef]
45. Colabardini, A.C.; Ries, L.N.A.; Brown, N.A.; Dos Reis, T.F.; Savoldi, M.; Goldman, M.H.S.; Menino, J.F.; Rodrigues, F.; Goldman, G.H. Functional characterization of a xylose transporter in Aspergillus nidulans. *Biotechnol. Biofuels* **2014**, *7*, 46. [CrossRef]
46. Wilcox, L.J.; Balderes, D.A.; Wharton, B.; Tinkelenberg, A.H.; Rao, G.; Sturley, S.L. Transcriptional profiling identifies two members of the ATP-binding cassette transporter superfamily required for sterol uptake in yeast. *J. Biol. Chem.* **2002**, *277*, 32466–32472. [CrossRef]
47. Bignell, E.; Cairns, T.C.; Throckmorton, K.; Nierman, W.C.; Keller, N.P. Secondary metabolite arsenal of an opportunistic pathogenic fungus. *Philos. Trans. R. Soc. B Biol. Sci.* **2016**, *371*, 20160023. [CrossRef]
48. Yin, W.B.; Baccile, J.A.; Bok, J.W.; Chen, Y.; Keller, N.P.; Schroeder, F.C. A nonribosomal peptide synthetase-derived iron(III) complex from the pathogenic fungus aspergillus fumigatus. *J. Am. Chem. Soc.* **2013**, *135*, 2064–2067. [CrossRef]
49. Baccile, J.A.; Spraker, J.E.; Le, H.H.; Brandenburger, E.; Gomez, C.; Woo Bok, J.; Macheleidt, J.; Brakhage, A.A.; Hoffmeister, D.; Keller, N.P.; et al. Plant-like biosynthesis of isoquinoline alkaloids in *Aspergillus fumigatus*. *Physiol. Behav.* **2016**, *176*, 139–148. [CrossRef]
50. Raffa, N.; Keller, N.P. A call to arms: Mustering secondary metabolites for success and survival of an opportunistic pathogen. *PLoS Pathog.* **2019**, *15*, e1007606. [CrossRef]
51. Horikoshi, R.; Goto, K.; Mitomi, M.; Oyama, K.; Sunazuka, T.; Omura, S. Identification of pyripyropene A as a promising insecticidal compound in a microbial metabolite screening. *J. Antibiot.* **2017**, *70*, 272–276. [CrossRef] [PubMed]
52. Zhan, Y.; Zhang, X.W.; Xiong, Y.; Li, B.L.; Nan, F.J. Design and synthesis of simple, yet potent and selective non-ring-A pyripyropene A-based inhibitors of acyl-coenzyme A: Cholesterol acyltransferase 2 (ACAT2). *Org. Biomol. Chem.* **2016**, *14*, 747–751. [CrossRef] [PubMed]
53. Chooi, Y.H.; Fang, J.; Liu, H.; Filler, S.G.; Wang, P.; Tang, Y. Genome mining of a prenylated and immunosuppressive polyketide from pathogenic fungi. *Org. Lett.* **2013**, *15*, 780–783. [CrossRef]
54. Brown, A.J.P.; Cowen, L.E.; Di Pietro, A.; Quinn, J. Stress Adaptation. *Fungal Kingd.* **2017**, *5*, 463–485. [CrossRef]
55. Hagiwara, D.; Sakamoto, K.; Abe, K.; Gomi, K. Signaling pathways for stress responses and adaptation in Aspergillus species: Stress biology in the post-genomic era. *Biosci. Biotechnol. Biochem.* **2016**, *80*, 1667–1680. [CrossRef] [PubMed]
56. Shenton, D.; Grant, C.M. Protein S-thiolation targets glycolysis and protein synthesis in response to oxidative stress in the yeast Saccharomyces cerevisiae. *Biochem. J.* **2003**, *374*, 513–519. [CrossRef] [PubMed]
57. Qiao, J.; Kontoyiannis, D.; Calderone, R.; Li, D.; Ma, Y.; Wan, Z.; Li, R.; Liu, W. Af yap1, encoding a bZip transcriptional factor of *Aspergillus fumigatus*, contributes to oxidative stress response but is not essential to the virulence of this pathogen in mice immunosuppressed by cyclophosphamide and triamcinolone. *Med. Mycol.* **2008**, *46*, 773–782. [CrossRef]

58. Lessing, F.; Kniemeyer, O.; Wozniok, I.; Loeffler, J.; Kurzai, O.; Haertl, A.; Brakhage, A.A. The *Aspergillus fumigatus* transcriptional regulator AfYap1 represents the major regulator for defense against reactive oxygen intermediates but is dispensable for pathogenicity in an intranasal mouse infection model. *Eukaryot. Cell* **2007**, *6*, 2290–2302. [CrossRef]
59. Richie, D.L.; Hartl, L.; Aimanianda, V.; Winters, M.S.; Fuller, K.K.; Miley, M.D.; White, S.; McCarthy, J.W.; Latgé, J.P.; Feldmesser, M.; et al. A role for the unfolded protein response (UPR) in virulence and antifungal susceptibility in *Aspergillus fumigatus*. *PLoS Pathog.* **2009**, *5*, e1000258. [CrossRef]
60. Ron, D.; Walter, P. Signal integration in the endoplasmic reticulum unfolded protein response. *Nat. Rev. Mol. Cell Biol.* **2007**, *8*, 519–529. [CrossRef]
61. Zhai, P.; Song, J.; Gao, L.; Lu, L. A sphingolipid synthesis-related protein OrmA in *Aspergillus fumigatus* is responsible for azole susceptibility and virulence. *Cell. Microbiol.* **2019**, *21*, e13092. [CrossRef]
62. Ruggiano, A.; Mora, G.; Buxó, L.; Carvalho, P. Spatial control of lipid droplet proteins by the ERAD ubiquitin ligase Doa10. *EMBO J.* **2016**, *35*, 1644–1655. [CrossRef]
63. Frawley, D.; Bayram, Ö. Identification of SkpA-CulA-F-box E3 ligase complexes in pathogenic Aspergilli. *Fungal Genet. Biol.* **2020**, *140*, 103396. [CrossRef]
64. Jöhnk, B.; Bayram, Ö.; Abelmann, A.; Heinekamp, T.; Mattern, D.J.; Brakhage, A.A.; Jacobsen, I.D.; Valerius, O.; Braus, G.H. SCF Ubiquitin Ligase F-box Protein Fbx15 Controls Nuclear Co-repressor Localization, Stress Response and Virulence of the Human Pathogen *Aspergillus fumigatus*. *PLoS Pathog.* **2016**, *12*, e1005899. [CrossRef] [PubMed]
65. Liu, Y.; Chang, A. Heat shock response relieves ER stress. *EMBO J.* **2008**, *27*, 1049–1059. [CrossRef] [PubMed]
66. Verghese, J.; Abrams, J.; Wang, Y.; Morano, K.A. Biology of the Heat Shock Response and Protein Chaperones: Budding Yeast (Saccharomyces cerevisiae) as a Model System. *Microbiol. Mol. Biol. Rev.* **2012**, *76*, 115–158. [CrossRef]
67. Meixner-monori, B.; Kubicek, C.P.; Habison, A.; Kubicek-pranz, E.V.A.M. Presence and Regulation of the ox-Ketoglutarate Dehydrogenase. *J. Bacteriol.* **1985**, *161*, 265–271. [CrossRef] [PubMed]
68. McLain, A.L.; Szweda, P.A.; Szweda, L.I. α-Ketoglutarate dehydrogenase: A mitochondrial redox sensor. *Free Radic. Res.* **2011**, *45*, 29–36. [CrossRef]
69. Albrecht, D.; Guthke, R.; Brakhage, A.A.; Kniemeyer, O. Integrative analysis of the heat shock response in *Aspergillus fumigatus*. *BMC Genom.* **2010**, *11*, 32. [CrossRef]
70. Sumegi, B.; Srere, P.A. Complex I binds several mitochondrial NAD-coupled dehydrogenases. *J. Biol. Chem.* **1984**, *259*, 15040–15045. [CrossRef]
71. Choi, J.K.; Kim, Y.J. Epigenetic regulation and the variability of gene expression. *Nat. Genet.* **2008**, *40*, 141–147. [CrossRef] [PubMed]
72. Landry, C.R.; Oh, J.; Hartl, D.L.; Cavalieri, D. Genome-wide scan reveals that genetic variation for transcriptional plasticity in yeast is biased towards multi-copy and dispensable genes. *Gene* **2006**, *366*, 343–351. [CrossRef] [PubMed]
73. Brown, A.J.P.; Larcombe, D.E.; Pradhan, A. Thoughts on the evolution of Core Environmental Responses in yeasts. *Fungal Biol.* **2020**, *124*, 475–481. [CrossRef] [PubMed]
74. Antal, K.; Gila, B.C.; Pócsi, I.; Emri, T. General stress response or adaptation to rapid growth in Aspergillus nidulans? *Fungal Biol.* **2020**, *124*, 376–386. [CrossRef]
75. Baltussen, T.J.H.; Coolen, J.P.M.; Zoll, J.; Verweij, P.E.; Melchers, W.J.G. Gene co-expression analysis identifies gene clusters associated with isotropic and polarized growth in *Aspergillus fumigatus* conidia. *Fungal Genet. Biol.* **2018**, *116*, 62–72. [CrossRef]
76. Baltussen, T.J.H.; Coolen, J.P.M.; Verweij, P.E.; Dijksterhuis, J.; Melchers, W.J.G. Identifying conserved generic aspergillus spp. Co-expressed gene modules associated with germination using cross-platform and cross-species transcriptomics. *J. Fungi* **2021**, *7*, 270. [CrossRef]
77. Losada, L.; Barker, B.M.; Pakala, S.; Pakala, S.; Joardar, V.; Zafar, N.; Mounaud, S.; Fedorova, N.; Nierman, W.C.; Cramer, R.A. Large-scale transcriptional response to hypoxia in *Aspergillus fumigatus* observed using RNAseq identifies a novel hypoxia regulated ncRNA. *Mycopathologia* **2014**, *178*, 331–339. [CrossRef]
78. Watkins, T.N.; Liu, H.; Chung, M.; Hazen, T.H.; Dunning Hotopp, J.C.; Filler, S.G.; Bruno, V.M. Comparative transcriptomics of *Aspergillus fumigatus* strains upon exposure to human airway epithelial cells. *Microb. Genom.* **2018**, *4*, e000154. [CrossRef]
79. Gibbons, J.G.; Beauvais, A.; Beau, R.; McGary, K.L.; Latgé, J.P.; Rokas, A. Global transcriptome changes underlying colony growth in the opportunistic human pathogen *Aspergillus fumigatus*. *Eukaryot. Cell* **2012**, *11*, 68–78. [CrossRef]
80. Remmele, C.W.; Luther, C.H.; Balkenhol, J.; Dandekar, T.; Müller, T.; Dittrich, M.T. Integrated inference and evaluation of host-fungi interaction networks. *Front. Microbiol.* **2015**, *6*, 764. [CrossRef]
81. Laothanachareon, T.; Asin-Garcia, E.; Volkers, R.J.M.; Tamayo-Ramos, J.A.; Martins Dos Santos, V.A.P.; Schaap, P.J. Identification of Aspergillus niger Aquaporins Involved in Hydrogen Peroxide Signaling. *J. Fungi* **2023**, *9*, 499. [CrossRef]
82. Everaert, C.; Luypaert, M.; Maag, J.L.V.; Cheng, Q.X.; DInger, M.E.; Hellemans, J.; Mestdagh, P. Benchmarking of RNA-sequencing analysis workflows using whole-transcriptome RT-qPCR expression data. *Sci. Rep.* **2017**, *7*, 1559. [CrossRef] [PubMed]
83. Hu, C.; Zhou, M.; Wang, W.; Sun, X.; Yarden, O.; Li, S. Abnormal ergosterol biosynthesis activates transcriptional responses to antifungal azoles. *Front. Microbiol.* **2018**, *9*, 9. [CrossRef]

84. Henry, K.W.; Nickels, J.T.; Edlind, T.D. Upregulation of ERG genes in Candida species by azoles and other sterol biosynthesis inhibitors. *Antimicrob. Agents Chemother.* **2000**, *44*, 2693–2700. [CrossRef] [PubMed]
85. Konecna, A.; Hervay, N.T.; Bencova, A.; Morvova, M.; Sikurova, L.; Jancikova, I.; Gaskova, D.; Gbelska, Y. Erg6 gene is essential for stress adaptation in Kluyveromyces lactis. *FEMS Microbiol. Lett.* **2018**, *365*, fny265. [CrossRef]
86. Ganz, T. Iron in Innate Immunity: Starve the Invaders Tomas. *Clin. Lymphoma* **2010**, *9*, 19–22. [CrossRef]
87. Paul, S.; Diekema, D.; Moye-Rowley, W.S. Contributions of *Aspergillus fumigatus* ATP-binding cassette transporter proteins to drug resistance and virulence. *Eukaryot. Cell* **2013**, *12*, 1619–1628. [CrossRef] [PubMed]
88. Esquivel, B.D.; Rybak, J.M.; Barker, K.S.; Fortwendel, J.R.; Rogers, P.D.; White, T.C. Characterization of the Efflux Capability and Substrate Specificity of *Aspergillus fumigatus* PDR5-like ABC Transporters Expressed in Saccharomyces cerevisiae. *mBio* **2020**, *11*, 1–16. [CrossRef] [PubMed]
89. Guillemette, T.; van Peij, N.N.M.E.; Goosen, T.; Lanthaler, K.; Robson, G.D.; van den Hondel, C.A.M.J.J.; Stam, H.; Archer, D.B. Genomic analysis of the secretion stress response in the enzyme-producing cell factory Aspergillus niger. *BMC Genom.* **2007**, *8*, 158. [CrossRef]
90. Coughlan, C.A.; Chotirmall, S.H.; Renwick, J.; Hassan, T.; Low, T.B.; Bergsson, G.; Eshwika, A.; Bennett, K.; Dunne, K.; Greene, C.M.; et al. The effect of *Aspergillus fumigatus* infection on vitamin D receptor expression in cystic fibrosis. *Am. J. Respir. Crit. Care Med.* **2012**, *186*, 999–1007. [CrossRef]
91. Macheleidt, J.; Scherlach, K.; Neuwirth, T.; Schmidt-Heck, W.; Straßburger, M.; Spraker, J.; Baccile, J.A.; Schroeder, F.C.; Keller, N.P.; Hertweck, C.; et al. Transcriptome analysis of cyclic AMP-dependent protein kinase A-regulated genes reveals the production of the novel natural compound fumipyrrole by *Aspergillus fumigatus*. *Mol. Microbiol.* **2015**, *96*, 148–162. [CrossRef] [PubMed]
92. Kurucz, V.; Krüger, T.; Antal, K.; Dietl, A.; Haas, H.; Pócsi, I. Additional oxidative stress reroutes the global response of *Aspergillus fumigatus* to iron depletion. *BMC Genom.* **2018**, *19*, 357. [CrossRef] [PubMed]
93. Bruns, S.; Seidler, M.; Albrecht, D.; Salvenmoser, S.; Remme, N.; Hertweck, C.; Brakhage, A.A.; Kniemeyer, O.; Müller, F.M.C. Functional genomic profiling of *Aspergillus fumigatus* biofilm reveals enhanced production of the mycotoxin gliotoxin. *Proteomics* **2010**, *10*, 3097–3107. [CrossRef]
94. Gaschler, M.M.; Stockwell, B.R. Lipid peroxidation in cell death. *Biochem. Biophys. Res. Commun.* **2017**, *482*, 419–425. [CrossRef] [PubMed]
95. Kobayashi, D.; Kondo, K.; Uehara, N.; Otokozawa, S.; Tsuji, N.; Yagihashi, A.; Watanabe, N. Endogenous Reactive Oxygen Species Is an Important Mediator of Miconazole Antifungal Effect. *Antimicrob. Agents Chemother.* **2002**, *46*, 3113–3117. [CrossRef] [PubMed]
96. Shekhova, E.; Kniemeyer, O.; Brakhage, A.A. Induction of Mitochondrial Reactive Oxygen Species Production by against *Aspergillus fumigatus*. *Antimicrob. Agents Chemother.* **2017**, *61*, e00978-17. [CrossRef]
97. Lambou, K.; Lamarre, C.; Beau, R.; Dufour, N.; Latge, J.P. Functional analysis of the superoxide dismutase family in *Aspergillus fumigatus*. *Mol. Microbiol.* **2010**, *75*, 910–923. [CrossRef]
98. Jung, K.W.; Kang, H.A.; Bahn, Y.S. Essential Roles of the Kar2/BiP Molecular Chaperone Downstream of the UPR Pathway in Cryptococcus neoformans. *PLoS ONE* **2013**, *8*, e58956. [CrossRef]
99. Hsu, C.L.; Prasad, R.; Blackman, C.; Ng, D.T.W. Endoplasmic reticulum stress regulation of the Kar2p/BiP chaperone alleviates proteotoxicity via dual degradation pathways. *Mol. Biol. Cell* **2012**, *23*, 630–641. [CrossRef]
100. Lackie, R.E.; Maciejewski, A.; Ostapchenko, V.G.; Marques-Lopes, J.; Choy, W.Y.; Duennwald, M.L.; Prado, V.F.; Prado, M.A.M. The Hsp70/Hsp90 chaperone machinery in neurodegenerative diseases. *Front. Neurosci.* **2017**, *11*, 254. [CrossRef]
101. Jacob, T.R.; Peres, N.T.A.; Martins, M.P.; Lang, E.A.S.; Sanches, P.R.; Rossi, A.; Martinez-Rossi, N.M. Heat shock protein 90 (Hsp90) as a molecular target for the development of novel drugs against the dermatophyte Trichophyton rubrum. *Front. Microbiol.* **2015**, *6*, 1241. [CrossRef]
102. Willger, S.D.; Cornish, E.J.; Chung, D.; Fleming, B.A.; Lehmann, M.M.; Puttikamonkul, S.; Cramer, R.A. Dsc orthologs are required for hypoxia adaptation, triazole drug responses, and fungal virulence in *Aspergillus fumigatus*. *Eukaryot. Cell* **2012**, *11*, 1557–1567. [CrossRef] [PubMed]
103. Foresti, O.; Ruggiano, A.; Hannibal-Bach, H.K.; Ejsing, C.S.; Carvalho, P. Sterol homeostasis requires regulated degradation of squalene monooxygenase by the ubiquitin ligase Doa10/Teb4. *eLife* **2013**, *2013*, e00953. [CrossRef] [PubMed]
104. Hou, J.; Tang, H.; Liu, Z.; Österlund, T.; Nielsen, J.; Petranovic, D. Management of the endoplasmic reticulum stress by activation of the heat shock response in yeast. *FEMS Yeast Res.* **2014**, *14*, 481–494. [CrossRef]
105. Caspeta, L.; Chen, Y.; Nielsen, J. Thermotolerant yeasts selected by adaptive evolution express heat stress response at 30 °C. *Sci. Rep.* **2016**, *6*, 27003. [CrossRef]
106. Pontecorvo, G.; Roper, J.A.; Chemmons, L.M.; Macdonald, K.D.; Bufton, A.W.J. The Genetics of Aspergillus nidulans. *Adv. Genet.* **1953**, *5*, 141–238. [CrossRef]
107. Hill, T.W.; Kafer, E. Improved protocols for Aspergillus minimal medium: Trace element and minimal medium salt stock solutions. *Fungal Genet. Rep.* **2001**, *48*, 20–21. [CrossRef]
108. Dobin, A.; Davis, C.A.; Schlesinger, F.; Drenkow, J.; Zaleski, C.; Jha, S.; Batut, P.; Chaisson, M.; Gingeras, T.R. STAR: Ultrafast universal RNA-seq aligner. *Bioinformatics* **2013**, *29*, 15–21. [CrossRef]

109. Love, M.I.; Huber, W.; Anders, S. Moderated estimation of fold change and dispersion for RNA-seq data with DESeq2. *Genome Biol.* **2014**, *15*, 550. [CrossRef]
110. Priebe, S.; Kreisel, C.; Horn, F.; Guthke, R.; Linde, J. FungiFun2: A comprehensive online resource for systematic analysis of gene lists from fungal species. *Bioinformatics* **2015**, *31*, 445–446. [CrossRef]
111. De Valk, H.A.; Meis, J.F.G.M.; Curfs, I.M.; Muehlethaler, K.; Mouton, J.W.; Corné, H.W. Use of a Novel Panel of Nine Short Tandem Repeats for Exact and High-Resolution Fingerprinting of *Aspergillus fumigatus* Isolates Use of a Novel Panel of Nine Short Tandem Repeats for Exact and High-Resolution Fingerprinting of *Aspergillus fumigatus* Isolates. *J. Clin. Microbiol.* **2005**, *43*, 4112–4120. [CrossRef] [PubMed]

Journal of
Fungi

MDPI

Article

Structural and Functional Alterations Caused by Aureobasidin A in Clinical Resistant Strains of *Candida* spp.

Rodrigo Rollin-Pinheiro [1,*,†], Daniel Clemente de Moraes [2,†], Brayan Bayona-Pacheco [2,3], Jose Alexandre da Rocha Curvelo [2], Giulia Maria Pires dos Santos-Freitas [1], Mariana Ingrid Dutra da Silva Xisto [1], Luana Pereira Borba-Santos [4], Sonia Rozental [4], Antonio Ferreira-Pereira [2] and Eliana Barreto-Bergter [1,*]

[1] Laboratório de Química Biológica de Microrganismos, Departamento de Microbiologia Geral, Instituto de Microbiologia Paulo de Góes, Universidade Federal do Rio de Janeiro (UFRJ), Rio de Janeiro 21941-902, Brazil; giuliapires89@gmail.com (G.M.P.d.S.-F.); marylanax@gmail.com (M.I.D.d.S.X.)

[2] Laboratório de Bioquímica Microbiana, Departamento de Microbiologia Geral, Instituto de Microbiologia Paulo de Góes, Universidade Federal do Rio de Janeiro (UFRJ), Rio de Janeiro 21941-902, Brazil; danielcmoraes@micro.ufrj.br (D.C.d.M.); bbayona@uninorte.edu.co (B.B.-P.); alexandrecurvelo@hotmail.com (J.A.d.R.C.); apereira@micro.ufrj.br (A.F.-P.)

[3] Departamento de Medicina, División Ciencias de la Salud, Universidad del Norte, Km 5, Vía Puerto Colombia, Área Metropolitana de Barranquilla, Barranquilla 081007, Colombia

[4] Programa de Biologia Celular e Parasitologia, Instituto de Biofísica Carlos Chagas Filho, Universidade Federal do Rio de Janeiro (UFRJ), Rio de Janeiro 21941-902, Brazil; luanaborba@biof.ufrj.br (L.P.B.-S.); rozental@biof.ufrj.br (S.R.)

* Correspondence: rodrigorollin@gmail.com (R.R.-P.); eliana.bergter@micro.ufrj.br (E.B.-B.)

† These authors contributed equally to this work.

Abstract: *Candida* species are one of the most concerning causative agents of fungal infections in humans. The treatment of invasive *Candida* infections is based on the use of fluconazole, but the emergence of resistant isolates has been an increasing concern which has led to the study of alternative drugs with antifungal activity. Sphingolipids have been considered a promising target due to their roles in fungal growth and virulence. Inhibitors of the sphingolipid biosynthetic pathway have been described to display antifungal properties, such as myriocin and aureobasidin A, which are active against resistant *Candida* isolates. In the present study, aureobasidin A did not display antibiofilm activity nor synergism with amphotericin B, but its combination with fluconazole was effective against *Candida* biofilms and protected the host in an in vivo infection model. Alterations in treated cells revealed increased oxidative stress, reduced mitochondrial membrane potential and chitin content, as well as altered morphology, enhanced DNA leakage and a greater susceptibility to sodium dodecyl sulphate (SDS). In addition, it seems to inhibit the efflux pump CaCdr2p. All these data contribute to elucidating the role of aureobasidin A on fungal cells, especially evidencing its promising use in clinical resistant isolates of *Candida* species.

Keywords: aureobasidin A; sphingolipids; *Candida*; membrane; fungal infection

1. Introduction

Candida species are one of the most concerning causative agents of fungal infections in humans. They are responsible for a wide spectrum of diseases, which range from superficial mycosis to life-threatening disseminated candidiasis [1]. In the context of invasive infections, the most frequent species are *Candida albicans* and *Nakaseomyces glabrata* (formerly named *Candida glabrata*) [2]. The mortality rates of invasive candidiasis can reach up to 60%, especially in patients presenting predisposing conditions, such as HIV infection, organ transplantation, cancer and diabetes [3]. In healthcare environments, *Candida* infections are the most common fungal disease associated with nosocomial acquired infections [4].

The treatment of invasive *Candida* infections was based on the use of fluconazole, an azole antifungal drug that blocks the enzyme lanosterol 14-α demethylase and, thus, disrupts ergosterol synthesis [5,6]. However, the emergence of resistant isolates has been an increasing concern, especially those presenting multi-drug resistance based on the expression of different types of transporters of over-expressing azole targets [7]. For this reason, recommendations for candidiasis treatment have been recently changed and echinocandins are now considered the first choice of drug, followed by azoles [8–11]. Alternative antifungal therapy is limited due to a variety of factors, which include the absence of a wide diversity of antifungal classes and high levels of toxicity and side effects presented by the antifungal drugs currently available in clinical settings, which lead to the need for studies of new antifungal approaches [12].

Besides the class of azoles, there are only three other groups of antifungal molecules available in healthcare units, which target only three fungal cell structures: polyenes, which directly bind to ergosterol found in the plasma membrane and cause the release of cytoplasmic content; echinocandins, which inhibit the enzyme β(1,3)-glucan synthase and affect cell wall synthesis; and fluoropyrimidine analogs, which block DNA synthesis [13]. For these reasons, the development of new antifungal options, as well as the discovery of alternative targets in fungal cells, is an urgent need. In this context, sphingolipids have been considered a promising candidate as a new target on fungal cells due to its role on fungal growth, virulence and cellular signaling [14,15]. Recently, the group of molecules acylhydrazones has been shown to block glucosylceramide synthesis in different fungal pathogens, such as *C. neoformans*, *C. albicans*, *A. fumigatus* and *Pneumocystis murina*, but does not affect the lipid metabolism in mammalian cells [16]. Inhibitors of sphingolipid biosynthesis have already been described, such as myriocin and aureobasidin A, which presented antifungal activity against a variety of pathogenic fungi, such as *Aspergillus* and *Candida* species including *Candida auris* [17–21]. Aureobasidin A acts by blocking the inositol phosphoryl ceramide (IPC) synthase, which mediates the synthesis of IPCs that are one of the most important sphingolipids produced by fungi [22]. However, the sphingolipid biosynthesis is a complex metabolic pathway that can be modulated by fungal cells, as has already been demonstrated by lipidomic analysis in *Candida* species [21]. In addition, the use of sphingolipid inhibitors in pathogenic filamentous fungi, such as *Scedosporium* species, has revealed that many intermediate molecules from the biosynthetic pathway are affected when cells are treated with different inhibitors [23].

More recently, our group evidenced the antifungal effect of these inhibitors in fluconazole-resistant isolates of *C. albicans* and *C. glabrata* and their synergistic effect with fluconazole [24]. For these reasons, it is important to understand which modifications occur in *Candida* cells when treated with aureobasidin A. The present study aimed to elucidate the effect of aureobasidin A on fluconazole-resistant clinical strains of *C. albicans* and *C. glabrata* that overexpress MDR1, CDR1 and ERG11, which are typical mechanisms of resistance for azole antifungals, focusing its effect on fungal lipids and the interaction with antifungal drugs that target an important fungal lipid, ergosterol, such as fluconazole and amphotericin B.

2. Materials and Methods

2.1. Cell Lineages and Reagents

A total of five *Candida* strains were used in this study. *C. albicans* ATCC 10231D-5 and *C. glabrata* ATCC 2001D-5 were used as standard (ATTCC©, Manassas, VA, USA). Three clinical isolates displaying resistance patterns to fluconazole were also evaluated, *C. glabrata* 109 strain, *C. albicans* 1114 strain (both from our laboratory collection) and *C. albicans* 12-99 strain, kindly provided by Theodore White from University of Missouri, USA. For all experiments, the strains were grown on Yeast Extract Peptone Dextrose (YPD) agar and transferred to YPD broth and incubated at 37 °C for 18 h under agitation.

Saccharomyces cerevisiae mutant strains were also used: AD/1234567, deleted from all genes that encode ABC transporters related to the MDR phenotype (Pdr5p, Yor1p, Snq2p, Pdr10p, Pdr11p and Ycf1p); CaCDR1p+, CaCDR2p+, CaMDR1p+ and CgCDR1p+, that overexpress CaCdr1p, CaCdr2p, CaMdr1p and CgCdr1p, respectively, and present the same gene deletions as AD/1234567. Those mutants were kindly provided by Dr. Richard Cannon and Dr. Brian Monk (University of Otago—New Zealand).

Aureobasidin A (Sigma-Aldrich, St. Louis, MO, USA) and fluconazole (University Pharmacy, UFJF, Juiz de Fora-MG, Brazil) were used in susceptibility and synergism tests.

2.2. Susceptibility of Forming and Preformed Biofilms to Aureobasidin A and Fluconazole

Biofilm formation was analyzed according to Hacioglu and colleagues [25]. Firstly, an adhesion step was performed by adding 200 µL from a standardized suspension of *Candida* cells (1×10^7/mL) to each well of a polystyrene microplate and incubated for 1.5 h at 37 °C. Then, the supernatant containing non-adherent cells was removed and RPMI 1640 medium supplemented with MOPS, 2% glucose and 20% fetal bovine serum (FBS, Gibco, Waltham, MA, USA) was added in the absence (positive control) or presence of aureobasidin A (4–0.125 µg/mL) and/or fluconazole (256–258 µg/mL). Adherent cells were then incubated for 48 h at 37 °C.

For the preformed biofilm assay, cells were grown as described above in the absence of the compounds. After 24 h of biofilm formation, the supernatant was removed and supplemented RPMI was added in the absence (positive control) or presence of aureobasidin A (4–0.125 µg/mL) and/or fluconazole (256–258 µg/mL). An additional incubation of 24 h at 37 °C was performed to evaluate the anti-biofilm activity.

Evaluation of both biofilm formation and preformed biofilms was carried out using three chromogenic assays as previously described [26,27]: crystal violet, safranin and XTT assays were used to analyze the overall biomass, extracellular matrix and metabolic activity, respectively.

2.3. Interaction of Aureobasidin A with Amphotericin B and Fluconazole

Interaction analysis of aureobasidin A with antifungals was performed using the combination with fluconazole against preformed biofilms and with amphotericin B against planktonic cells. Both experiments were performed using the checkerboard method according to Reis de Sá and colleagues [28]. For the analysis against preformed biofilms, after 24 h of biofilm formation in the absence of the compounds, the supernatant was removed and supplemented RPMI was added in the absence (positive control) or presence of aureobasidin A (4–0.125 µg/mL) and/or fluconazole (256–258 µg/mL) in different directions. An additional incubation of 24 h at 37 °C was performed to evaluate the anti-biofilm activity of different combinations of aureobasidin A and fluconazole. For the combination of aureobasidin A and amphotericin B against planktonic cells, *Candida* cells were treated with both drugs using the checkerboard method and the results were evaluated after a 48 h incubation at 37 °C.

The fractional inhibitory index (FICI) was calculated according to the formula (MIC combined/MIC drug A alone) + (MIC combined/MIC drug B alone), where A is aureobasidin A and drug B is fluconazole. Interaction was classified according to the following parameter: $\leq$0.5, synergistic interaction; >0.5 to $\leq$4, no interaction; >4, antagonistic effect [29].

Bliss independence model was performed according to Meletiadis and colleagues and Zhao and colleagues [30,31]. The following formula was used to assess the drug interaction: $Eexp = Ea + Eb - Ea \times Eb$, in which Eexp is the expected efficacy of drug combination, Ea is the efficacy of drug A (aureobasidin A) and Eb is the efficacy of drug B (fluconazole). The results were classified as: synergistic effect, Eobs > Eexp; indifference, Eobs = Eexp; antagonistic effect, Eobs < Eexp.

2.4. Analyses of Cellular Alterations

Alterations of *Candida* cells caused by aureobasidin A were analyzed using fluorescent staining. Oxidative stress, mitochondrial membrane potential and the content of chitin and neutral lipids were evaluated using 2′,7′–dichlorofluorescein diacetate (DCFH-DA)

(Sigma-Aldrich, St. Louis, MO, USA), JC-1 probe (ThermoFisher, Waltham, MA, USA), calcofluor white (Sigma-Aldrich, St. Louis, MO, USA) and Nile Red (Sigma-Aldrich, St. Louis, MO, USA), respectively. Cells were grown in the absence (positive control) or in the presence of 0.5× MIC of aureobasidin A for 48 h at 37 °C. Then, cells were stained for 1 h at 37 °C under protection from light with 50 μg/mL of DCFH-DA, 10 μg/mL of JC-1, 25 μg/mL of calcofluor white or 2 mg/mL of Nile Red [32,33]. Samples were washed three times to remove residual dye and suspended in PBS. Fluorescence intensity was measured using the SpectraMax 340 microplate reader (Molecular Devices, San Jose, CA, USA) at the following conditions: DCFH-DA at 492 nm (excitation) and 517 nm (emission); JC-1 at 475 nm (excitation) and 529 nm (green fluorescence) or 590 nm (red fluorescence) for the calculation of the red/green fluorescence intensity; calcofluor white at 350 nm (excitation) and 432 nm (emission); Nile Red at 550 nm (excitation) and 635 nm (emission).

2.5. Susceptibility to SDS

The susceptibility to sodium dodecyl sulphate (SDS) was analyzed using 1×10^5 cells grown in 96-well plates containing supplemented RPMI in the absence (positive control) and in the presence of sub-inhibitory concentrations of aureobasidin A (0.5× MIC) for 24 h at 37 °C [23]. Then, the supernatant was removed from the microplates and SDS (90 μg/mL) was added. After another 24 h incubation, the cell viability was measured using an XTT reduction assay and readings were captured using a spectrophotometer (Bio-Rad, Hercules, CA, USA) at 490 nm.

2.6. Evaluation of Cell Membrane Permeability

The cell membrane permeability was determined by measuring the release of DNA to the culture supernatant [34]. *Candida* cells (1×10^5) were grown in supplemented RPMI in the absence (positive control) or presence of aureobasidin A (0.5× MIC) for 48 h at 37 °C. After the incubation time, the cells were pelleted and the supernatant analyzed using a NanoDrop 2000 spectrophotometer (Thermo Fisher Scientific) to quantify free DNA (260 nm). A sterile culture medium sample was used as a negative control.

2.7. Efflux Pump Inhibition

The ability of aureobasidin A in inhibiting efflux pumps was evaluated through checkerboard assays. Briefly, *S. cerevisiae* cells (2×10^4/mL) were incubated in YPD broth in the presence of combinations of 2-fold serial dilutions of aureobasidin A (1.25–0.078 μg/mL) and fluconazole (500–1.95 μg/mL) at 30 °C for 48 h. Cell growth was measured using a 600 nm SpectraMax 340 microplate reader (Molecular Devices, San Jose, CA, USA), and interactions were assessed through a FICI model, as described in Section 2.3.

2.8. Scanning Electron Microscopy

Scanning electron microscopy was performed according to Rollin-Pinheiro et al. [35]. *Candida* cells were grown in supplemented RPMI in the absence (positive control) or in the presence of 0.5× MIC of aureobasidin A and incubated for 48 h at 37 °C. After the period of incubation, the cell growth was collected and processed for scanning electron microscopy. Briefly, samples were processed as follows: i. fixation in 2.5% glutaraldehyde and 4% formaldehyde, in 0.1 M cacodylate buffer, for 30 min at room temperature; ii. post-fixation in 1% osmium tetroxide in 0.1 M cacodylate buffer containing 1.25% potassium ferrocyanide for 30 min; iii. dehydration in a graded ethanol series (30–100%); iv. critical point drying in CO_2 (EM CPD300, Leica, German); v. adhesion to aluminum stubs with carbon tape; and vi. coating with gold.

Images were obtained with FEI Quanta 250 scanning electron microscope (FEI Company, Hillsboro, OR, USA) and processed using Photoshop v.CS6 software (Adobe, San Jose, CA, USA). The yeast area was measured using the Image J v.1.52a software.

2.9. Caenorhabidtis Elegans Infection Assay

The in vivo infection assay used AU-37 *C. elegans* strain that was acquired by the *Caenorhabidtis* Genetics Center (CGC), University of Minnesota (USA). The worms were maintained in nematoid growth agar medium (NGM) (3 g/L of NaCl; 17 g/L of agar; 2.5 g/L of peptone; 1 mM $CaCl_2$; 5 mg/L of cholesterol in ethanol; 1 mM $MgSO_4$ and 25 mM KPO_4) at 15 °C, seeded with the strain of *Escherichia coli* OP50 that was used as the regular food source for these nematodes.

To assess the influence of 0.25 µg/mL aureobasidin A associated or not with 128 µg/mL of fluconazole on nematode infection caused by the three strains independently, Priya's [36] methodology was used with minor modifications, as follows. A synchronized population of nematodes was obtained by collecting eggs from pregnant adult larvae with M9 buffer (3 g of KH_2PO_4, 6 g of Na_2HPO_4, 5 g of NaCl, 1 mL of 1 M $MgSO_4$ in 1 L of H_2O) and vigorously vortexed with 0.5 mL of 5 M NaOH and 1.5 mL of 6% sodium hypochlorite in 35 s sequential cycles. Released eggs were then centrifuged at 1300× *g* for 50 s, cleaned in sequential washes in M9, added to plates containing NGM medium seeded with *E. coli* OP50 and incubated for 72 h at 25 °C to obtain synchronized larvae in stage L4 [37]. Infection of synchronized larvae with *C. albicans* 12-99, *C. albicans* 1114 and *C. glabrata* 109 was carried out as proposed by Tempakakis and colleagues [38], with minor modifications. Briefly, 5×10^5 cells of each strain were incubated in Petri dishes containing BHI agar supplemented with 90 µg/mL of kanamycin, 200 µg/mL of streptomycin and 200 µg/mL of ampicillin for 24 h at 37 °C. The plates were cooled to at room temperature and approximately 100 synchronized worms in stage L4 were added onto each plate seeded with different *Candida* strains and allowed to feed for 180 min at 25 °C and allowed to crawl on agar surface with the aim of feeding on yeast. Infected worms were then washed serially in M9 buffer and placed on clean BHI plates to crawl for 40 min, aiming to remove yeast from their cuticles [39]. Nematodes were collected from plates and resuspended in screen medium (60% M9, 40% BHI, 90 µg/mL kanamycin, 200 µg/mL streptomycin and 200 µg/mL ampicillin). To evaluate the influence of aureobasidin associated or not with fluconazole on nematode infection, 20 infected worms were incubated in 96-well TPP® plates (Schaffhausen, Switzerland) containing screen medium with 0.25 µg/mL aureobasidin alone or associated with 128 µg/mL of fluconazole and incubated at 25 °C for nine days maximum without shaking and scored as live or dead on daily basis. The survival of nematodes was monitored in an inverted microscope Axiovert100 Zeiss® (Jena, Germany) and determined through motility, sinus morphology and/or bulbopharyngeal movement of each helminth. The survival index was calculated from the proportion of the viable population in relation to the total number of worms, including live and dead animals.

2.10. Statistical Analyses

All experiments were performed in triplicate in three independent experimental sets. Statistical analyses were performed using GraphPad Prism v5.00 for Windows (GraphPad Software, San Diego, CA, USA). The nonparametric Kruskal–Wallis one-way analysis of variance was used to compare the observed differences among the groups, and individual comparisons of the groups were performed using a Bonferroni post-test. The 90% or 95% confidence interval was determined in all experiments.

For the in vivo analysis, the experiment was conducted independently two-fold, with a quadruplicate analysis in each one. Data were analyzed by Student's paired *t*-test, and p values lower than 0.05 were considered significant.

3. Results

3.1. Interaction between Aureobasidin A with Amphotericin B

In a previous study from our group it has been demonstrated that aureobasidin A displays synergistic effects with fluconazole in planktonic cells. For this reason, it was evaluated whether it also has synergism with amphotericin B, another antifungal agent used in clinical settings for the treatment of candidiasis. For the three *Candida* strains used, the interaction between aureobasidin A and amphotericin B was indifferent, indicating that there was no synergistic or antagonistic effect between the compounds (Table 1).

Table 1. Checkerboard assays of aureobasidin A combined with amphotericin B against *Candida* spp. clinical isolates.

Strains	Aureobasidin (μg/mL)			Amphotericin B (μg/mL)			FICI	Outcome
	MIC a	MIC c	FIC	MIC a	MIC c	FIC		
12-99	>0.25	>0.25	1	0.25	0.25	1	2	I
1114	>0.25	>0.25	1	0.25	0.25	1	2	I
109	>0.25	>0.25	1	0.25	0.25	1	2	I

MIC a, MIC of compound alone; MIC c, MIC of compound combined; FIC, fractional inhibitory concentration; FICI, fractional inhibitory concentration index; I, indifferent.

3.2. *Effect of Aureobasidin A against Candida Biofilms*

In previous studies from our group, it has been demonstrated that aureobasidin A is active against *Candida* planktonic cells, including clinical resistant isolates (109, 1114 and 12-99). *Candida glabrata* 109 and *Candida albicans* 1114 overexpress the CDR1 (ATP transporter) and MDR1 (MFS transporter) genes, respectively, whereas *Candida albicans* 12-99 overexpress the ERG11, CDR1, CDR2 and MDR1 genes [24].

The present study aimed to investigate the effect of aureobasidin A against biofilms produced by these three strains. The evaluation of forming biofilm revealed that it did not affect fungal biomass (Figure 1A) but reduced approximately 50% of the extracellular matrix at 2 and 4 μg/mL in strain 109 (Figure 1B). A decrease of at least 50% of biofilm viability was also observed in strain 109 at 2 and 4 μg/mL and in the strain 12-99 at 1, 2 and 4 μg/mL (Figure 1C). In strain 1114, aureobasidin A reduced biofilm viability at 4 μg/mL to 60% compared to the control (Figure 1C).

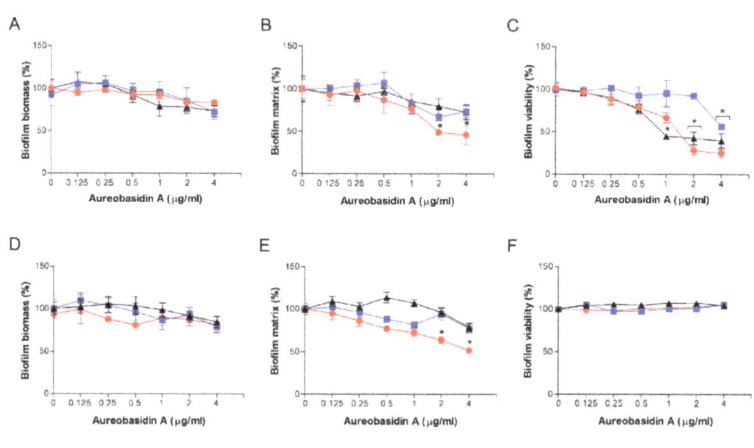

Figure 1. Effect of aureobasidin A on forming (**A–C**) and preformed (**D–F**) *Candida* biofilms. 109: *C. glabrata* 109; 1114: *C. albicans* 1114; 12-99: *C. albicans* 12-99. Fungal biomass (**A**,**D**), extracellular matrix (**B**,**E**) and viability (**C**,**F**) were measured using violet crystal, safranin and XTT reduction assay, respectively. * $p < 0.05$.

Regarding the preformed biofilm, when biofilm was formed prior to the treatment with aureobasidin A, it did not affect fungal biomass or viability in any *Candida* isolates (Figure 1D,F), but a reduction in the extracellular matrix was observed in strain 109 at 2 and 4 μg/mL (Figure 1E).

3.3. Effect of Aureobasidin A Combined with Fluconazole against Candida Biofilms

Considering that the interaction between aureobasidin A and fluconazole displayed synergism in planktonic cells, as demonstrated in previous a study [24], it was analyzed whether the same effect is observed when *Candida* cells are grown as biofilms. Using the FIC index method, there was no synergistic effect between aureobasidin A and fluconazole for any of the three *Candida* isolates (Table 2). However, when the data were analyzed by the BLISS method, a synergistic effect was observed in the strains 1114 and 12-99 (Table 3).

Table 2. Anti-biofilm activity of aureobasidin A and fluconazole—alone or in combination—using the Fractional Inhibitory Concentration Index (FICI) against *C. glabrata* 109, *C. albicans* 1114 and *C. albicans* 12-99.

Strains	Aureobasidin (μg/mL)			Fluconazole (μg/mL)			FICI	Outcome
	MIC a	MIC c	FIC	MIC a	MIC c	FIC		
12-99	>4	2	0.5	>256	64	0.25	0.75	I
1114	>4	2	0.5	>256	64	0.25	0.75	I
109	>4	2	0.5	>256	128	0.5	1.0	I

MIC a, MIC of compound alone; MIC c, MIC of compound combined; FIC, fractional inhibitory concentration; FICI, fractional inhibitory concentration index; I, indifferent.

Table 3. Anti-biofilm activity of aureobasidin A and fluconazole—alone or in combination—using the Bliss Independence method against *C. glabrata* 109, *C. albicans* 1114 and *C. albicans* 12-99.

Efficacy of Combined Drugs					
Strains	Drugs	MIC_{50} (μg/mL)	E_{obs}	E_{exp}	ΔE, % (Interaction)
109	Fluconazole Aureobasidin A	>256 >4.0	3.1	4.0	−0.9 (A)
1114	Fluconazole Aureobasidin A	>256 >4.0	10.0	4.1	5.9 (S)
12-99	Fluconazole Aureobasidin A	>256 >2.0	33.3	4.2	29.1 (S)

MIC, Minimal inhibitory concentration. E_{obs}, efficacy observed in the analysis. E_{exp}, efficacy expected according to Bliss calculation. ΔE, difference between E_{obs} and E_{exp}.

3.4. Alterations of Cellular Parameters Caused by Aureobasidin A

Since aureobasidin A displayed significant antifungal activity, it was investigated which alterations are observed in treated *Candida* cells. DCFDA, JC-1, calcofluor white and Nile Red were used to evaluate the oxidative stress, depolarization of mitochondrial membrane, chitin and neutral lipids content, respectively. Treatment with aureobasidin A led to an increase in ROS production in both the 109, 1114 and 12-99 strains between 1.5 and 2.0 fold compared to the control, suggesting that the compound resulted in oxidative stress (Figure 2A). JC-1 staining was at least 50% decreased in the treated cells, indicating that aureobasidin A caused depolarization of the mitochondrial membrane (Figure 2B). Regarding the analyses of molecules from the fungal surface, the chitin content measured by calcofluor white was around 30–40% reduced by aureobasidin A treatment (Figure 2C), whereas the staining of neutral lipids by Nile Red was not altered in the treated cells (Figure 2D).

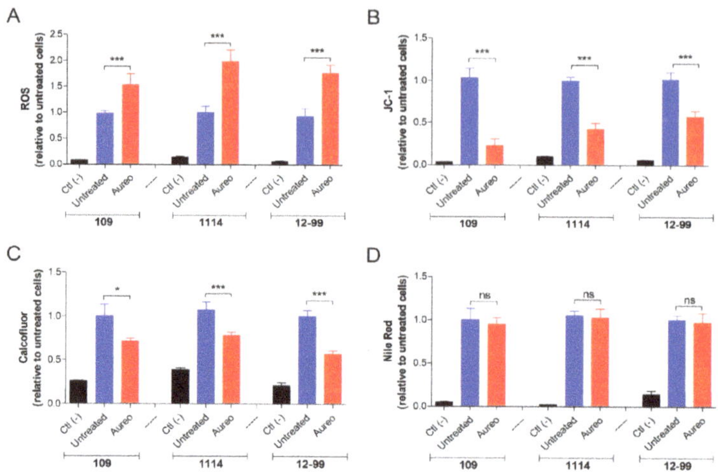

Figure 2. Cellular alterations caused by aureobasidin A on *C. glabrata* 109, *C. albicans* 1114 and *C. albicans* 12-99 after incubation with the compound for 48 h at 37 °C. Oxidative stress, mitochondrial membrane potential, chitin and neutral lipid contents were measured using the fluorescent probes DCFDA, JC-1, calcofluor white and Nile Red, respectively. * $p < 0.05$; *** $p < 0.001$. ns = non significative.

3.5. Scanning Electron Microscopy

Since the treatment of *Candida* cells with aurebasidin A revealed some cellular alterations, modifications on fungal morphology were evaluated by scanning electron microscopy. Compared to the control of untreated cells, which showed isolated yeast cells, all the three strains presented grouped cells when incubated with aureobasidin A (Figure 3). In addition, the area of yeast cells was measured and it revealed that the cell size of *C. glabrata* 109 was reduced when treated with aurebasidin A, whereas it was increased in *C. albicans* 1114 and 12-99 (Figure 3).

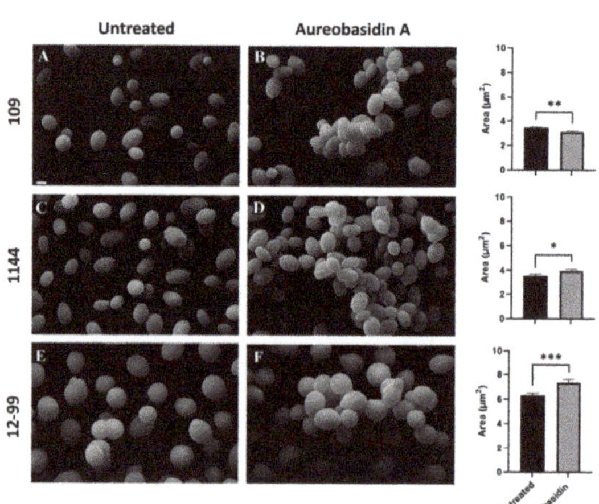

Figure 3. Scanning electron microscopy of *C. glabrata* 109, *C. albicans* 1114 and *C. albicans* 12-99 grown for 48 h at 37 °C in the presence or the absence of aureobasidin A. Cell area was measured using the software Image J v.1.52a. Untreated (**A**,**C**,**E**): strains 109, 1114 and 12-99, respectively, grown in the absence of aureobasidin A. Aureobasidin A (**B**,**D**,**F**): strains 109, 1114 and 12-99, respectively, grown in the presence of the compound. * $p < 0.05$; ** $p < 0.01$; *** $p < 0.001$.

3.6. Susceptibility to SDS in the Presence of Aureobasidin A

Considering that aureobasidin A inhibits the synthesis of an important class of lipids found on fungal membrane, sodium dodecyl sulphate (SDS) was used as a stressor of plasma membrane to check whether the compound could increase the susceptibility of *Candida* cells to SDS. To do this, a subinhibitory concentration of SDS was used to check whether the treatment with aureobasidin A increased the susceptibility of *Candida* cells to the stressor. The viability of all the three strains was reduced when treated with 1/2 MIC of aureobasidin A, suggesting that it enhances the susceptibility of fungal cells to SDS (Figure 4).

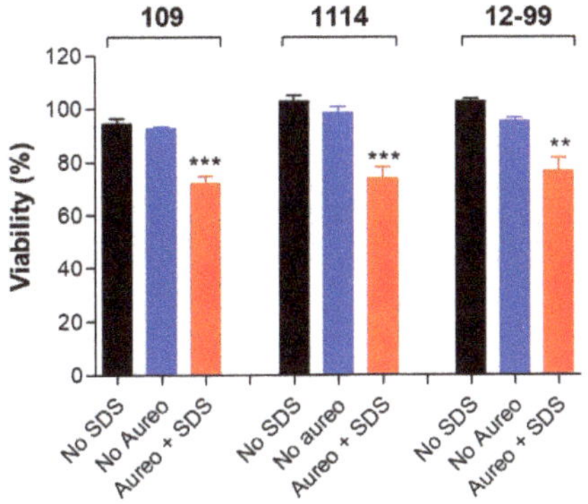

Figure 4. Susceptibility of *C. glabrata* 109, *C. albicans* 1114 and *C. albicans* 12-99 to aureobasidin A in the presence of the membrane stressor SDS. Cells were grown for 48 h at 37 °C in the absence or the presence of ½ MIC of aureobasidin A and a subinhibitory concentration of SDS. Cell viability was assessed using the XTT reduction assay. No SDS: cells grown in the absence of both SDS and aureobasidin A. No Aureo: cells grown in the presence of SDS and the absence of aureobasidin A. Aureo + SDS: cells grown in the presence of both SDS and aureobasidin A. ** $p < 0.01$; *** $p < 0.001$.

3.7. Membrane Permeability in the Presence of Aureobasidin A

Considering that the treatment with aureobasidin A resulted in more susceptible cells, the permeability of fungal membrane in the presence of the compound was investigated by measuring the DNA concentration found in the supernatant of culture medium. Whereas the control cells displayed around 20 ng/µL of DNA, the treatment with aureobasidin A resulted in an increase in DNA content in the supernatant of 30–40 ng/µL (Figure 5), suggesting that the compound could induce the permeability of fungal membranes.

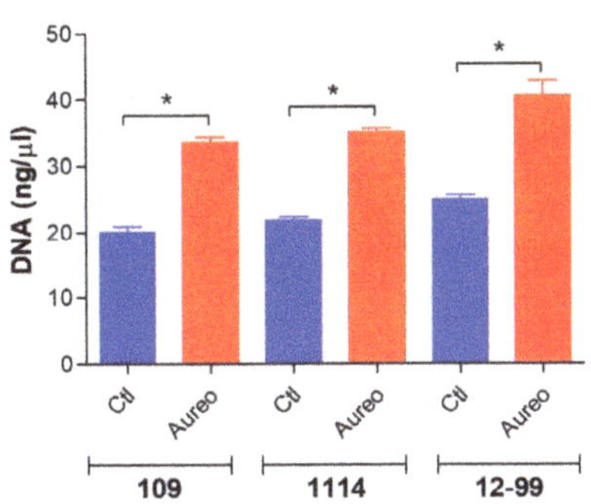

Figure 5. Analysis of DNA content in the supernatant of fungal growth. *C. glabrata* 109, *C. albicans* 1114 and *C. albicans* 12-99 were grown for 48 h at 37 °C in the absence (CTL) or the presence (AUREO) of aureobasidin A. After centrifugation, supernatant was collected and DNA content was measured by NanoDrop 2000 spectrophotometer at 260 nm. * $p < 0.001$.

3.8. Efflux Pump Inhibition

Checkerboard assays performed with *S. cerevisiae* mutant strains show that aureobasidin A displays antifungal activity against *S. cerevisiae*, with MIC of 1.25 µg/mL for the five strains tested. Fluconazole presented 0.5 µg/mL MIC against AD/1234567, 125 µg/mL against CaCDR2p+ and CaMDR1p+ strains, and 500 µg/mL against CaCDR1p+ and CgCDR1p+ strains. Combining aureobasidin A with fluconazole led to FICI values of 0.52–2.00 against the AD/1234567, CaCDR1p+, CaMDR1p+ and CgCDR1p+ strains, indicating indifferent interactions. The association between the compounds against CaCDR2p+ resulted in a synergistic interaction, with a FICI value of 0.28 (Table 4).

Table 4. Checkerboard assays of aureobasidin A against *S. cerevisiae* mutant strains overexpressing *Candida* spp. efflux pumps.

Strains	Aureobasidin (µg/mL)			Fluconazole (µg/mL)			FICI	Outcome
	MIC a	MIC c	FIC	MIC a	MIC c	FIC		
AD/1234567	1.25	1.25	1.00	0.5	0.5	1.00	2.00	I
CaCDR1p+	1.25	0.63	0.50	500	125	0.25	0.75	I
CaCDR2p+	1.25	0.31	0.25	125	3.91	0.03	0.28	S
CaMDR1p+	1.25	0.63	0.50	125	2.00	0.02	0.52	I
CgCDR1p+	1.25	0.63	0.50	500	125	0.25	0.75	I

MIC a, MIC of compound alone; MIC c, MIC of compound combined; FIC, fractional inhibitory concentration; FICI, fractional inhibitory concentration index; I, indifferent; S, synergistic.

3.9. In Vivo Assay

Based on previous results that evaluated if aureobasidin A was synergic in vitro with fluconazole, enhancing its effect on different resistant *Candida* strains with specific resistant patterns [24], aureobasidin A was also tested in vivo in a *C. elegans* model of *Candida* infection to evaluate the behavior of the aureobasidin A/fluconazole interaction as a treatment to counteract *Candida* infection in the immunocompromised *C. elegans* AU-37. Briefly, worms were infected with each *Candida* strain and then treated with 0.25 µg/mL of aureobasidin A, associated or not with 128 µg/mL fluconazole.

As presented in Figure 6, more than 47.3% of the worms infected with *C. glabrata* 109 died after one day and only 5% were alive after seven days. A very similar result was observed for worms treated with 128 µg/mL fluconazole, with more than 55% of worms dead one day after treatment and more than 90% of worms dead by the seventh day post-infection. Infected worms treated with 0.25 µg/mL of aureobasidin A exhibited a slight but non-significant prolonged survival. However, when worms were treated with fluconazole and aureobasidin A combined, infection was significantly reduced in all times tested. Nematodes survival reached 91.5, 66.6, 46.5, 43.8, 38.8, 30.4 and 27.5% for 24, 48, 72, 96, 12, 144 and 168 h post infection, respectively (Figure 6A).

The infection caused by the *C. albicans* 1114 strain demonstrated a controversial result that is represented in Figure 6B. Such infection evolved in a more stable manner but reached no significant efficacy in any treatment test until the fifth post-treatment day, where the survival rate significantly reached 51.7, 50.7 and 50.8% for combined treatment against 44.1, 36.8 and 26.2% for untreated infected nematodes on the fifth, sixth and seventh day, respectively (Figure 6B).

Lastly, the infection caused by the *C. albicans* 12-99 strain, although it evolved in a more stable manner as well, revealed a treatment efficacy resembling the one exposed by the infection caused by strain 109 (Figure 6C). The treatment with the association of aureobasidin A and fluconazole showed a significant survival rate compared with the untreated worms for all times tested. This rate was 91.2, 86.1, 82, 73.2, 67.5, 60.8, 52.1% against 73.5, 63.6, 63.3, 53.9, 47.3, 36.4 and 25.8% for the first to the seventh day, respectively. Nevertheless, the treatment with 0.25 µg/mL of aureobasidin A alone was also able to significantly enhance worms' survival in the first four post-infection days compared with the untreated cells. This survival percentage was calculated as 90.4, 73.9, 72.1 and 63.3% for the first to the fourth day, respectively (Figure 6C).

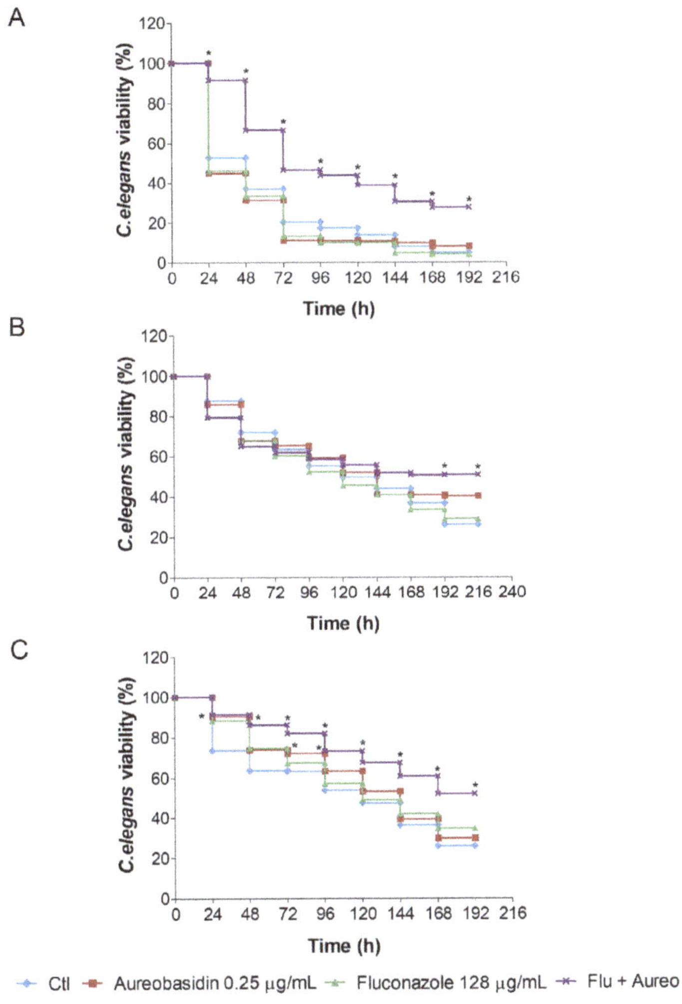

Figure 6. In vivo assay using the *C. elegans* model of infection. After infection by *C. glabrata* 109 (**A**), *C. albicans* 1114 (**B**) or *C. albicans* 12-99 (**C**), worms were treated with aureobasidin A, fluconazole or the combination aureobasidin A/fluconazole (Flu + Aureo). CTL: infected worms grown with no treatment regimen. *C. elegans* viability was measured at every 24 h interval. * $p < 0.001$.

4. Discussion

A recent study from our group has demonstrated that aureobasidin A is active against resistant clinical isolates of *Candida* species (109, 1114 and 12-99) and displays a synergistic effect with fluconazole [24]. Nevertheless, there are no studies showing what occurs with fungal cells when sphingolipid synthesis is impaired by inhibitors.

Since aureobasidin A is active against planktonic cells, as seen in the previous study, its effect on *Candida* biofilms was evaluated. Whereas it was active against biofilm formation, especially reducing the extracellular matrix and biofilm viability, a less prominent effect was observed against preformed biofilms. Tan and Tay [19] have also tested aurebasidin

A against *Candida* biofilms and it has been shown that its susceptibility moved from 1 and 4 µg/mL in *C. glabrata* and *C. albicans* planktonic cells, respectively, to >64 µg/mL in biofilms of both species. Similar results have been observed for other *Candida* species [19]. These data indicate that fungal biofilms are more resistant to aureobasidin A, which has already been observed for many other antifungal molecules.

Considering that aureobasidin A and fluconazole presented synergistic effects [24], the interaction between aureobasidin A and amphotericin B was checked but no synergistic effect was observed. On the other hand, we also evaluated the combination of aureobasidin A and fluconazole against biofilms. Although no synergism was seen by the FIC index, BLISS method revealed a promising positive interaction between the drugs, especially for strain 12-99. Analyses of drug interaction involving the use of aureobasidin are scarce in the literature; therefore, more studies are needed to elucidate how this inhibitor interacts with the antifungal agents currently used in clinical settings.

Although it is well known that aureobasidin A acts on sphingolipid biosynthesis, there are few data in the literature showing modifications caused in fungal treated cells. Aureobasidin A MIC in *Cryptococcus neoformans* was not affected in the presence of sorbitol, suggesting that the compound does not interfere with the synthesis of cell wall components [40]. However, our data showed a decrease in chitin content measured using calcofluor white, suggesting that disruption of the sphingolipid biosynthetic pathway may result in modifications of the structure of the cell wall. On the other hand, it has been shown that the expression of the Cxt1p gene in *C. neoformans*, which is responsible for the synthesis of galactoxylomannan of the fungal capsule, is decreased when cells are treated with aureobasidin A [40]. It suggests that the biosynthesis of other polysaccharides could also be affected by the compound, which would corroborate the reduction in chitin content observed in *Candida* cells in our data; but, more studies are needed to elucidate the effect of aureobasidin A on cell wall components.

We found that *Candida* cells displayed increased oxidative stress and decreased mitochondrial membrane potential after aureobasidin A treatment, suggesting that the disruption of sphingolipid biosynthesis may also result in mitochondrial damage. In *C. neoformans*, it has also been shown that aureobasidin A treatment led to the production of oxidative stress, probably because of mitochondrial membrane disturbance and a reduction in lysosomal integrity [40]. It might indicate that the synthesis of sphingolipids is crucial for the integrity of fungal mitochondria.

Scanning electron microscopy revealed that treatment with aureobasidin A resulted in grouped cells compared to the control. Morphological alterations have also been demonstrated in *C. neoformans* by transmission electron microscopy, in which cells displayed excessive vacuolization and deposition of lipid globules [40].

Considering that aureobasidin A inhibits the synthesis of sphingolipids, which compose fungal membranes, it is reasonable to investigate in more detail the modifications it causes in *Candida* membrane integrity. Aureobasidin A increased the susceptibility of treated cells to SDS and induced the release of DNA to the supernatant, suggesting that sphingolipids might play a role in membrane integrity and susceptibility to detergents. Similar results were found for other eukaryotic cells, such as *Saccharomyces cerevisiae*, *C. neoformans* and *Toxoplasma gondii*, in which crushed vacuoles, leakage of ions K^+ and the loss of intracellular structures, respectively, were observed [40–42]. All these data indicate that the inhibition of sphingolipid causes a variety of consequences on cellular membranes.

In *S. cerevisiae*, deletion of *ERG6* leads to resistance to aureobasidin A, indicating that alterations in genes responsible for ergosterol biosynthesis impact the susceptibility to sphingolipid inhibitors [43]. Interestingly, the *C. albicans* 12-99 strain, which overexpresses *ERG11*, is susceptible to aureobasidin A, suggesting that other genes from the ergosterol biosynthetic pathway are not related to the susceptibility to sphingolipid inhibitors. These data demonstrate that ergosterol and sphingolipid biosynthesis can be correlated.

Mishra and Prasad [44] have described the lipid distribution in *C. albicans* cells. Phospholipids and free fatty acids are the major lipids found in plasma membrane, followed by ergosterol and sphingolipids. Lipids are also present in the fungal cell wall, including ergosterol, fatty acid, phospholipids and sphingolipids, and its presence in the cell wall has been suggested to regulate *C. albicans'* adhesion to host cells [44]. These data could help to explain why cells treated with aurebasidin A seem to be linked to each other, as observed by scanning electron microscopy, but more studies are needed to elucidate it.

To evaluate whether aureobasidin A affects efflux pumps from *C. albicans* and *C. glabrata*, *S. cerevisiae* mutant strains that overexpress *Candida* MDR transporters were used. Since each *S. cerevisiae* strain harbors one specific efflux pump, it is possible to assess exactly which one is inhibited by a drug. Data showed that aureobasidin A inhibits CaCdr2p, due to the synergism observed. Moreover, combining aureobasidin A with fluconazole reduced the MIC of the antifungal by 64 times against CaMDR1p+, indicating that CaMdr1p is also inhibited by the same compound. It would explain the synergistic interactions between aureobasidin A and fluconazole against 1114, a *C. albicans* isolate that overexpresses CaMdr1p, and 12-99, a strain that overexpresses CaCdr1p, CaCdr2p and CaMdr1p [24]. Moreover, aureobasidin A displayed the same antifungal activity against all the *S. cerevisiae* strains tested in this study, indicating that this substance is not a substrate for the transporters. Therefore, it may be presumed that: (1) efflux inhibition by aureobasidin A is not related to competitive inhibition at the catalytic site of the transporters; and (2) *Candida* spp. would not become resistant to aureobasidin A over time. Corroborating our data, it has already been demonstrated that the transporter Cdr1p is localized in enriched ergosterol and sphingolipid microdomains [45], and the deletion of the *IPT1* (inositol phosphotransferase) gene decreases the level of Cdr1p on plasma membranes [46].

The in vivo efficacy of aureobasidin A was assessed using the *C. elegans* model of infection. For all of the three strains, the combination of aureobasidin A with fluconazole was effective in increasing the worm's survival, corroborating the fact that aureobasidin A and fluconazole display in vitro synergism against *Candida* species. In vivo studies investigating the antifungal activity of aureobasidin A are rare in the literature. Takesako and colleagues [47] have performed a murine model of candidiasis and have demonstrated that the treatment with aureobasidin A protected mice against *Candida* infections. In addition, a previous study has shown that aureobasidin A is not toxic for *C. elegans* [24], indicating that its low toxicity is displayed not only in in vitro cell cultures, but also in in vivo analyses. These data suggest that it is a promising effective drug in the context of the disease established in a host.

All the data presented here bring a more detailed analysis of how aureobasidin A affects *Candida* cells, especially clinical resistant strains. Nevertheless, further studies are needed to test more *Candida* isolates and species, elucidate how sphingolipid biosynthesis is modulated by aureobasidin A and clarify how it interacts with antifungal agents. On the other hand, the evidence of an alternative drug with low MIC values, non-toxicity properties and in vivo efficacy to treat resistant isolates is a promising area to be explored to improve the management of non-susceptible invasive candidiasis in clinical settings.

Author Contributions: Conceptualization, R.R.-P., D.C.d.M. and B.B.-P.; methodology, R.R.-P., D.C.d.M., B.B.-P., J.A.d.R.C., G.M.P.d.S.-F., M.I.D.d.S.X. and L.P.B.-S.; software, R.R.-P., D.C.d.M. and B.B.-P.; validation, R.R.-P., D.C.d.M. and B.B.-P.; formal analysis, R.R.-P., D.C.d.M. and B.B.-P.; investigation, R.R.-P., D.C.d.M., B.B.-P., J.A.d.R.C., G.M.P.d.S.-F., M.I.D.d.S.X. and L.P.B.-S.; resources, S.R., A.F.-P. and E.B.-B.; data curation, R.R.-P., D.C.d.M. and B.B.-P.; writing—original draft preparation, R.R.-P., D.C.d.M. and J.A.d.R.C.; writing—review and editing, A.F.-P. and E.B.-B.; supervision, A.F.-P. and E.B.-B.; project administration, A.F.-P. and E.B.-B.; funding acquisition, A.F.-P. and E.B.-B. All authors have read and agreed to the published version of the manuscript.

Funding: This study was financed in part by the Coordenação de Aperfeiçoamento de Pessoal de Nível Superior—Brasil (CAPES)—Finance Code 001; Conselho Nacional de Desenvolvimento Científico e TecnolÓgico (CNPq) Universal-Processo # 408981/2018-0 and Fundação de Amparo à Pesquisa do estado do Rio de Janeiro-CNE (Faperj) Processo # E-26/202974/2017. Xisto, M.I.D.S. was supported by FAPERJ #E-26/202.051/2020, Borba-Santos, L.P. was supported by FAPERJ #E-26/203.853/2022, Freitas, G.M.P.S. was supported by #E-26/202.088/2022 and Rollin-Pinheiro, R.

was supported by FAPERJ #E-26/200.591/2022 and #E-26/210.355/2022. The funders had no role in study design, data collection and analysis, decision to publish or preparation of the manuscript. Funding support was provided to buy reagents for the development of the study.

Institutional Review Board Statement: Not applicable.

Informed Consent Statement: Not applicable.

Data Availability Statement: Data are contained within the article.

Acknowledgments: The authors thank the Centro Nacional de Biologia Estrutural e Bioimagem (CENABIO, UFRJ, Rio de Janeiro, Brazil) for the facilities and support in the use of SEM equipment.

Conflicts of Interest: The authors declare no conflict of interest.

References

1. Poulain, D. *Candida albicans*, plasticity and pathogenesis. *Crit. Rev. Microbiol.* **2015**, *41*, 208–217. [CrossRef] [PubMed]
2. Dadar, M.; Tiwari, R.; Karthik, K.; Chakraborty, S.; Shahali, Y.; Dhama, K. *Candida albicans*—Biology, molecular characterization, pathogenicity, and advances in diagnosis and control—An update. *Microb. Pathog.* **2018**, *117*, 128–138. [CrossRef] [PubMed]
3. Singh, A.; Rella, A.; Schwacke, J.; Vacchi-Suzzi, C.; Luberto, C.; Del Poeta, M. Transmembrane transporter expression regulated by the glucosylceramide pathway in *Cryptococcus neoformans*. *BMC Res. Notes* **2015**, *8*, 681. [CrossRef] [PubMed]
4. Bongomin, F.; Gago, S.; Oladele, R.O.; Denning, D.W. Global and Multi-National Prevalence of Fungal Diseases-Estimate Precision. *J. Fungi* **2017**, *3*, 57. [CrossRef]
5. Rajendran, R.; Sherry, L.; Nile, C.J.; Sherriff, A.; Johnson, E.M.; Hanson, M.F.; Williams, C.; Munro, C.A.; Jones, B.J.; Ramage, G. Biofilm formation is a risk factor for mortality in patients with *Candida albicans* bloodstream infection-Scotland, 2012–2013. *Clin. Microbiol. Infect.* **2016**, *22*, 87–93. [CrossRef] [PubMed]
6. Pristov, K.E.; Ghannoum, M.A. Resistance of Candida to azoles and echinocandins worldwide. *Clin. Microbiol. Infect.* **2019**, *25*, 792–798. [CrossRef]
7. Arendrup, M.C.; Patterson, T.F. Multidrug-Resistant Candida: Epidemiology, Molecular Mechanisms, and Treatment. *J. Infect. Dis.* **2017**, *216*, S445–S451. [CrossRef]
8. Pappas, P.G.; Lionakis, M.S.; Arendrup, M.C.; Ostrosky-Zeichner, L.; Kullberg, B.J. Invasive candidiasis. *Nat. Rev. Dis. Primers* **2018**, *4*, 18026. [CrossRef]
9. von Lilienfeld-Toal, M.; Wagener, J.; Einsele, H.; Cornely, O.A.; Kurzai, O. Invasive Fungal Infection. *Dtsch. Arztebl. Int.* **2019**, *116*, 271–278. [CrossRef]
10. Bassetti, M.; Righi, E.; Montravers, P.; Cornely, O.A. What has changed in the treatment of invasive candidiasis? A look at the past 10 years and ahead. *J. Antimicrob. Chemother.* **2018**, *73*, i14–i25. [CrossRef]
11. Pappas, P.G.; Kauffman, C.A.; Andes, D.R.; Clancy, C.J.; Marr, K.A.; Ostrosky-Zeichner, L.; Reboli, A.C.; Schuster, M.G.; Vazquez, J.A.; Walsh, T.J.; et al. Executive Summary: Clinical Practice Guideline for the Management of Candidiasis: 2016 Update by the Infectious Diseases Society of America. *Clin. Infect. Dis.* **2016**, *62*, 409–417. [CrossRef]
12. Roemer, T.; Krysan, D.J. Antifungal drug development: Challenges, unmet clinical needs, and new approaches. *Cold Spring Harb. Perspect. Med.* **2014**, *4*, a019703. [CrossRef] [PubMed]
13. Campoy, S.; Adrio, J.L. Antifungals. *Biochem. Pharmacol.* **2017**, *133*, 86–96. [CrossRef]
14. Del Poeta, M.; Nimrichter, L.; Rodrigues, M.L.; Luberto, C. Synthesis and biological properties of fungal glucosylceramide. *PLoS Pathog.* **2014**, *10*, e1003832. [CrossRef] [PubMed]
15. Mota Fernandes, C.; Del Poeta, M. Fungal sphingolipids: Role in the regulation of virulence and potential as targets for future antifungal therapies. *Expert. Rev. Anti-Infect. Ther.* **2020**, *18*, 1083–1092. [CrossRef] [PubMed]
16. Lazzarini, C.; Haranahalli, K.; Rieger, R.; Ananthula, H.K.; Desai, P.B.; Ashbaugh, A.; Linke, M.J.; Cushion, M.T.; Ruzsicska, B.; Haley, J.; et al. Acylhydrazones as Antifungal Agents Targeting the Synthesis of Fungal Sphingolipids. *Antimicrob. Agents Chemother.* **2018**, *62*, e00156. [CrossRef] [PubMed]
17. Cerantola, V.; Guillas, I.; Roubaty, C.; Vionnet, C.; Uldry, D.; Knudsen, J.; Conzelmann, A. Aureobasidin A arrests growth of yeast cells through both ceramide intoxication and deprivation of essential inositolphosphorylceramides. *Mol. Microbiol.* **2009**, *71*, 1523–1537. [CrossRef]
18. Lattif, A.A.; Mukherjee, P.K.; Chandra, J.; Roth, M.R.; Welti, R.; Rouabhia, M.; Ghannoum, M.A. Lipidomics of *Candida albicans* biofilms reveals phase-dependent production of phospholipid molecular classes and role for lipid rafts in biofilm formation. *Microbiology* **2011**, *157*, 3232–3242. [CrossRef]
19. Tan, H.W.; Tay, S.T. The inhibitory effects of aureobasidin A on Candida planktonic and biofilm cells. *Mycoses* **2013**, *56*, 150–156. [CrossRef]
20. Perdoni, F.; Signorelli, P.; Cirasola, D.; Caretti, A.; Galimberti, V.; Biggiogera, M.; Gasco, P.; Musicanti, C.; Morace, G.; Borghi, E. Antifungal activity of Myriocin on clinically relevant Aspergillus fumigatus strains producing biofilm. *BMC Microbiol.* **2015**, *15*, 248. [CrossRef]

21. Kumar, M.; Singh, A.; Kumari, S.; Kumar, P.; Wasi, M.; Mondal, A.K.; Rudramurthy, S.M.; Chakrabarti, A.; Gaur, N.A.; Gow, N.A.R.; et al. Sphingolipidomics of drug resistant Candida auris clinical isolates reveal distinct sphingolipid species signatures. *Biochim. Biophys. Acta Mol. Cell Biol. Lipids* **2021**, *1866*, 158815. [CrossRef] [PubMed]
22. Aeed, P.A.; Young, C.L.; Nagiec, M.M.; Elhammer, A.P. Inhibition of inositol phosphorylceramide synthase by the cyclic peptide aureobasidin A. *Antimicrob. Agents Chemother.* **2009**, *53*, 496–504. [CrossRef] [PubMed]
23. Rollin-Pinheiro, R.; Rochetti, V.P.; Xisto, M.; Liporagi-Lopes, L.C.; Bastos, B.; Rella, A.; Singh, A.; Rozental, S.; Del Poeta, M.; Barreto-Bergter, E. Sphingolipid biosynthetic pathway is crucial for growth, biofilm formation and membrane integrity of Scedosporium boydii. *Future Med. Chem.* **2019**, *11*, 2905–2917. [CrossRef] [PubMed]
24. Rollin-Pinheiro, R.; Bayona-Pacheco, B.; Domingos, L.T.S.; da Rocha Curvelo, J.A.; de Castro, G.M.M.; Barreto-Bergter, E.; Ferreira-Pereira, A. Sphingolipid Inhibitors as an Alternative to Treat Candidiasis Caused by Fluconazole-Resistant Strains. *Pathogens* **2021**, *10*, 856. [CrossRef]
25. Hacioglu, M.; Oyardi, O.; Kirinti, A. Oregano essential oil inhibits *Candida* spp. biofilms. *Z. Naturforsch. C J. Biosci.* **2021**, *76*, 443–450. [CrossRef]
26. Mello, T.P.; Aor, A.C.; Gonçalves, D.S.; Seabra, S.H.; Branquinha, M.H.; Santos, A.L. Assessment of biofilm formation by *Scedosporium apiospermum, S. aurantiacum, S. minutisporum* and *Lomentospora prolificans*. *Biofouling* **2016**, *32*, 737–749. [CrossRef]
27. Mello, T.P.; Oliveira, S.S.C.; Frasés, S.; Branquinha, M.H.; Santos, A.L.S. Surface properties, adhesion and biofilm formation on different surfaces by *Scedosporium* spp. and *Lomentospora prolificans*. *Biofouling* **2018**, *34*, 800–814. [CrossRef]
28. Reis de Sá, L.F.; Toledo, F.T.; de Sousa, B.A.; Gonçalves, A.C.; Tessis, A.C.; Wendler, E.P.; Comasseto, J.V.; Dos Santos, A.A.; Ferreira-Pereira, A. Synthetic organotelluride compounds induce the reversal of Pdr5p mediated fluconazole resistance in Saccharomyces cerevisiae. *BMC Microbiol.* **2014**, *14*, 201. [CrossRef]
29. Odds, F.C. Synergy, antagonism, and what the chequerboard puts between them. *J. Antimicrob. Chemother.* **2003**, *52*, 1. [CrossRef]
30. Meletiadis, J.; Petraitis, V.; Petraitiene, R.; Lin, P.; Stergiopoulou, T.; Kelaher, A.M.; Sein, T.; Schaufele, R.L.; Bacher, J.; Walsh, T.J. Triazole-polyene antagonism in experimental invasive pulmonary aspergillosis: In vitro and in vivo correlation. *J. Infect. Dis.* **2006**, *194*, 1008–1018. [CrossRef] [PubMed]
31. Zhao, W.; Sachsenmeier, K.; Zhang, L.; Sult, E.; Hollingsworth, R.E.; Yang, H. A New Bliss Independence Model to Analyze Drug Combination Data. *J. Biomol. Screen.* **2014**, *19*, 817–821. [CrossRef] [PubMed]
32. Rollin-Pinheiro, R.; Almeida, Y.C.; Rochetti, V.P.; Xisto, M.; Borba-Santos, L.P.; Rozental, S.; Barreto-Bergter, E. Miltefosine Against *Scedosporium* and *Lomentospora* Species: Antifungal Activity and Its Effects on Fungal Cells. *Front. Cell. Infect. Microbiol.* **2021**, *11*, 698662. [CrossRef]
33. de Oliveira, E.B.; Xisto, M.; Rollin-Pinheiro, R.; Rochetti, V.P.; Barreto-Bergter, E. Peptidorhamnomannans from *Scedosporium* and *Lomentospora* Species Display Microbicidal Activity Against Bacteria Commonly Present in Cystic Fibrosis Patients. *Front. Cell. Infect. Microbiol.* **2020**, *10*, 598823. [CrossRef]
34. Spadari, C.C.; Vila, T.; Rozental, S.; Ishida, K. Miltefosine Has a Postantifungal Effect and Induces Apoptosis in Cryptococcus Yeasts. *Antimicrob. Agents Chemother.* **2018**, *62*, e00312. [CrossRef]
35. Rollin-Pinheiro, R.; Xisto, M.; de Castro-Almeida, Y.; Rochetti, V.P.; Borba-Santos, L.P.; Fontes, Y.D.S.; Ferreira-Pereira, A.; Rozental, S.; Barreto-Bergter, E. Pandemic Response Box® library as a source of antifungal drugs against *Scedosporium* and *Lomentospora* species. *PLoS ONE* **2023**, *18*, e0280964. [CrossRef] [PubMed]
36. Priya, A.; Pandian, S.K. Piperine Impedes Biofilm Formation and Hyphal Morphogenesis of *Candida albicans*. *Front. Microbiol.* **2020**, *11*, 756. [CrossRef]
37. Swetha, T.K.; Pooranachithra, M.; Subramenium, G.A.; Divya, V.; Balamurugan, K.; Pandian, S.K. Umbelliferone Impedes Biofilm Formation and Virulence of Methicillin-Resistant *Staphylococcus epidermidis* via Impairment of Initial Attachment and Intercellular Adhesion. *Front. Cell. Infect. Microbiol.* **2019**, *9*, 357. [CrossRef] [PubMed]
38. Tampakakis, E.; Okoli, I.; Mylonakis, E. A *C. elegans*-based, whole animal, in vivo screen for the identification of antifungal compounds. *Nat. Protoc.* **2008**, *3*, 1925–1931. [CrossRef]
39. Breger, J.; Fuchs, B.B.; Aperis, G.; Moy, T.I.; Ausubel, F.M.; Mylonakis, E. Antifungal chemical compounds identified using a C. elegans pathogenicity assay. *PLoS Pathog.* **2007**, *3*, e18. [CrossRef] [PubMed]
40. Teymuri, M.; Shams-Ghahfarokhi, M.; Razzaghi-Abyaneh, M. Inhibitory effects and mechanism of antifungal action of the natural cyclic depsipeptide, aureobasidin A against *Cryptococcus neoformans*. *Bioorganic Med. Chem. Lett.* **2021**, *41*, 128013. [CrossRef]
41. Zhen, C.; Lu, H.; Jiang, Y. Novel Promising Antifungal Target Proteins for Conquering Invasive Fungal Infections. *Front. Microbiol.* **2022**, *13*, 911322. [CrossRef] [PubMed]
42. Sonda, S.; Sala, G.; Ghidoni, R.; Hemphill, A.; Pieters, J. Inhibitory effect of aureobasidin A on Toxoplasma gondii. *Antimicrob. Agents Chemother.* **2005**, *49*, 1794–1801. [CrossRef] [PubMed]
43. Fukuda, S.; Kono, Y.; Ishibashi, Y.; Tabuchi, M.; Tani, M. Impaired biosynthesis of ergosterol confers resistance to complex sphingolipid biosynthesis inhibitor aureobasidin A in a PDR16-dependent manner. *Sci. Rep.* **2023**, *13*, 11179. [CrossRef]
44. Mishra, P.; Prasad, R. An overview of lipids of *Candida albicans*. *Prog. Lipid Res.* **1990**, *29*, 65–85. [CrossRef]
45. Urbanek, A.K.; Muraszko, J.; Derkacz, D.; Łukaszewicz, M.; Bernat, P.; Krasowska, A. The Role of Ergosterol and Sphingolipids in the Localization and Activity of *Candida albicans*' Multidrug Transporter Cdr1p and Plasma Membrane ATPase Pma1p. *Int. J. Mol. Sci.* **2022**, *23*, 9975. [CrossRef]

46. Mukhopadhyay, K.; Prasad, T.; Saini, P.; Pucadyil, T.J.; Chattopadhyay, A.; Prasad, R. Membrane sphingolipid-ergosterol interactions are important determinants of multidrug resistance in *Candida albicans*. *Antimicrob. Agents Chemother.* **2004**, *48*, 1778–1787. [CrossRef] [PubMed]
47. Takesako, K.; Kuroda, H.; Inoue, T.; Haruna, F.; Yoshikawa, Y.; Kato, I.; Uchida, K.; Hiratani, T.; Yamaguchi, H. Biological properties of aureobasidin A, a cyclic depsipeptide antifungal antibiotic. *J. Antibiot.* **1993**, *46*, 1414–1420. [CrossRef] [PubMed]

Journal of *Fungi*

MDPI

Article

Miltefosine: A Repurposing Drug against Mucorales Pathogens

Mariana Ingrid Dutra da Silva Xisto [1,*,†], Rodrigo Rollin-Pinheiro [1,†], Victor Pereira Rochetti [1], Yuri de Castro-Almeida [1], Luana Pereira Borba-Santos [2], Giulia Maria Pires dos Santos-Freitas [1], Jefferson Cypriano [3], Fernanda de Ávila Abreu [3], Sonia Rozental [2] and Eliana Barreto-Bergter [1,*]

[1] Laboratório de Química Biológica de Microrganismos, Departamento de Microbiologia Geral, Instituto de Microbiologia Paulo de Góes, Universidade Federal do Rio de Janeiro, Rio de Janeiro 21941-902, Brazil; rodrigorollin@gmail.com (R.R.-P.); victorrochetti@gmail.com (V.P.R.); yuricastro20155@gmail.com (Y.d.C.-A.); giuliapires89@gmail.com (G.M.P.d.S.-F.)

[2] Laboratório de Biologia Celular de Fungos, Programa de Biologia Celular e Parasitologia, Instituto de Biofísica Carlos Chagas Filho, Universidade Federal do Rio de Janeiro, Rio de Janeiro 21941-902, Brazil; luanaborba@biof.ufrj.br (L.P.B.-S.); rozental@biof.ufrj.br (S.R.)

[3] Laboratório de Biologia Celular e Magnetotaxia & Unidade de Microscopia Multiusuário, Instituto de Microbiologia Paulo de Góes, Universidade Federal do Rio de Janeiro, Rio de Janeiro 21941-902, Brazil; jeffcy@micro.ufrj.br (J.C.); fernandaaabreu@micro.ufrj.br (F.d.Á.A.)

* Correspondence: marylanax@gmail.com (M.I.D.d.S.X.); eliana.bergter@micro.ufrj.br (E.B.-B.)

† These authors contributed equally to this work.

Abstract: Mucorales are a group of non-septated filamentous fungi widely distributed in nature, frequently associated with human infections, and are intrinsically resistant to many antifungal drugs. For these reasons, there is an urgent need to improve the clinical management of mucormycosis. Miltefosine, which is a phospholipid analogue of alkylphosphocholine, has been considered a promising repurposing drug to be used to treat fungal infections. In the present study, miltefosine displayed antifungal activity against a variety of Mucorales species, and it was also active against biofilms formed by these fungi. Treatment with miltefosine revealed modifications of cell wall components, neutral lipids, mitochondrial membrane potential, cell morphology, and the induction of oxidative stress. Treated Mucorales cells also presented an increased susceptibility to SDS. Purified ergosterol and glucosylceramide added to the culture medium increased miltefosine MIC, suggesting its interaction with fungal lipids. These data contribute to elucidating the effect of a promising drug repurposed to act against some relevant fungal pathogens that significantly impact public health.

Keywords: miltefosine; Mucorales; *Rhizopus oryzae*; *Rhizopus stolonifer*; *Rhizopus microsporus*; *Cunninghamella* spp.; *Mucor velutinosus*; drug repurposing

1. Introduction

Mucorales are a group of non-septated filamentous fungi widely distributed in nature and usually found in decomposing organic matter or soil [1]. The species most frequently found in human infections are *Rhizopus*, *Lichtheimia* (formerly known as *Absidia* or *Mycocladus*), and *Mucor* [2]. Mucormycosis is an invasive fungal infection that presents high mortality rates due to its aggressive pattern, difficulties in its diagnosis, and intrinsic high resistance to many antifungals used in clinical settings [3]. Some risk factors are associated with the establishment of mucormycosis, such as diabetes, chemotherapy, organ transplantation, hematological diseases, and high serum levels of iron [4]. Mucorales are considered opportunistic pathogens because they are frequently related to infections in immunocompromised patients, and although their incidence is low, cases have been increasing during the last few decades, especially in this group of patients [4]. The most common clinical manifestations are rhino-cerebral, pulmonary, cutaneous, gastrointestinal, and disseminated infections [5]. Angioinvasion is an important characteristic of invasive mucormycosis, which consists of penetration of the endothelium and the formation of thrombosis and tissue necrosis, facilitating the hematological dissemination of the fungi [1].

A concerning factor that contributes to the bad prognosis of mucormycosis is its intrinsic resistance to many antifungal drugs, such as amphotericin B, posaconazole, and isavuconazole. Interestingly, these are still the recommended options to treat mucormycosis in clinical settings, indicating that the development of alternative antifungal therapies is an urgent need [6]. In addition, many cases of Mucorales infections have been recently associated with COVID-19 as a post-COVID syndrome, which highlights the necessity of improving the clinical management of these mycoses [7–10].

In this context, miltefosine has been seen as a promising repurposing drug for the treatment of fungal infections. It is a phospholipid analog from the alkyl phosphocholine class that was first developed as an antitumor agent, but studies have already shown its activity against a variety of microorganisms, including *Leishmania* spp. and *Trypanosoma cruzi* [11–13]. Its antifungal activity has been demonstrated in dermatophytes, *Cryptococcus* spp., *Candida* spp., *Sporothrix* spp., *Paracoccidioides* spp., *Histoplasma capsulatum*, *Coccidioides posadasii*, *Rhizopus* spp., *Aspergillus* spp., *Fusarium* spp., and *Scedosporium* spp. [14–22].

Little is known about the mechanism of action of miltefosine, especially in Mucorales. Some studies have already shown that it affects ergosterol biosynthesis and increases plasma membrane permeability in dimorphic fungi and *Scedosporium* species [15,23]. In addition, it seems to directly interact with ergosterol and glucosylceramide on fungal membranes, which could help to explain the disruption to the plasma membrane caused by miltefosine [23,24]. Since the exact mechanisms of action of miltefosine have not been clarified yet, the present study aims to evaluate the effects caused by this drug on different species of Mucorales. The null hypothesis was that miltefosine did not show significant difference in terms of cell alterations between different Mucorales species tested. In addition, miltefosine did not exhibit fungicidal activity in concentrations of up to 64 µg/mL against the Mucorales species, as was observed in several other fungi.

2. Materials and Methods

2.1. Microorganisms

Rhizopus oryzae UCP1295, *Rhizopus stolonifera* UCP1300, and *Rhizopus microsporus* var. *microsporus* UCP1304 isolated from the Brazilian Caatinga area were supplied by Galba Maria de Campos-Takaki from the Culture Collection (RENNORFUN - Rede Norte Nordeste de Fungos Filamentosos) from the Catholic University of Pernambuco, Recife, Brazil. Fungal stocks were kept in potato dextrose medium. Clinical strains of *Mucor velutinosus* H136BO and *Cunninghamella* spp. B926 were supplied by Marcio Nucci from the Mycology Laboratory of the University Hospital, Universidade Federal do Rio de Janeiro. To obtain conidia, cells were grown on potato dextrose agar plates for seven days at room temperature. Conidia were obtained by washing the plate surface with phosphate-buffered saline (PBS, pH 7.2), and hyphal fragments and debris were removed by filtration through a cell strainer (Falcon). The suspension was then centrifuged and cells were counted in Neubauer's chamber to be used in the experiments. These five Mucorales species were used in all experiments of this study.

2.2. Antifungal Susceptibility Testing

The susceptibility of Mucorales species to miltefosine was determined according to EUCAST protocols, with modifications [25], using the broth microdilution method. A stock solution of miltefosine (Cayman Chemical Co., Ann Arbor, MI, USA) was prepared in dimethyl sulphoxide:ethanol (1:1) and maintained at −20 °C. To evaluate the minimum inhibitory concentration, miltefosine was serially diluted (0.062–64 µg/mL) and added to 96-well plates in the presence of a standardized suspension of conidia (3×10^4), followed by an incubation in RPMI 1640 medium (Sigma-Aldrich, St. Louis, MO, USA). After 72 h at 37 °C in 5% CO_2, cell growth was analyzed by OD readings at 600 nm in a spectrophotometer (Bio-Rad, Hercules, CA, USA). Cell viability was assessed using the XTT-reduction assay. The minimum inhibitory concentration (MIC_{70}) of miltefosine was defined as the lowest concentration that inhibits 70% of fungal growth.

After the MIC analyses, to determine the minimum fungicidal concentration (MFC), aliquots of 10 µL of each well were plated on potato dextrose agar and incubated at room temperature for 3 days.

2.3. Anti-Biofilm Assays

Miltefosine activity was evaluated against Mucorales mature biofilms. Firstly, biofilms were formed in 96-well plates for 24 h [26,27]. Briefly, a conidia suspension of each species (1×10^5) was added to each well and incubated for 1.5 h at 37 °C in 5% CO_2 for the adhesion stage. Further, the supernatant was removed to discard non-adherent cells, and fresh RPMI 1640 medium was added for biofilm formation for 24 h at 37 °C. After that, the medium was removed and miltefosine (1/4 MIC_{70}–8 MIC_{70}) was added. Positive control consisted of biofilm formation in the absence of miltefosine. An additional incubation for 24 h at 37 °C in 5% CO_2 was performed to evaluate miltefosine activity. Biofilms were analyzed by three parameters [26,27]: crystal violet for overall biomass, safranin for extracellular matrix, and XTT-reduction assay for metabolic activity.

2.4. Susceptibility to SDS and NaCl

To evaluate the susceptibility to sodium dodecyl sulfate (SDS) and NaCl, conidia of Mucorales species (1×10^5) were grown in 96-well plates containing RPMI in the presence of miltefosine (1/2 MIC_{70} or MIC_{70}) and co-incubated with either SDS (sub-inhibitory concentration of 45 µg/mL for *Rhizopus* species and *M. velutinosus* and 90 µg/mL for *Cunninghamella* spp.) or 1% NaCl. Positive control consisted of cells grown in the absence of miltefosine. After 72 h incubation at 37 °C in 5% CO_2, the cell viability was measured by using the XTT-reduction assay and readings were captured using a spectrophotometer (Bio-Rad, Hercules, CA, USA) at 490 nm.

2.5. Susceptibility in the Presence of Exogenous Ergosterol and Glucosylceramide

Mucorales susceptibility to miltefosine in the presence of exogenous ergosterol (Sigma-Aldrich, St. Louis, MO, USA) or GlcCer (purified from *R. stolonifer* UCP1300) was determined according to EUCAST protocols, with modifications [24,25]. Conidia of Mucorales species (3×10^4) were incubated in the presence of miltefosine in serial dilutions as described above in the antifungal susceptibility test section. Whereas fungi incubated only with miltefosine were used as a control of MIC_{70} values, cells were also grown with miltefosine using RPMI supplemented with purified ergosterol (50 and 100 µg/mL) and GlcCer (50 and 100 µg/mL) to check their susceptibility to miltefosine in the presence of these lipids [23]. After 72 h incubation at 37 °C in 5% CO_2, cell growth was measured in a spectrophotometer at 600 nm (Bio-Rad, Hercules, CA, USA) and cell viability was determined by the XTT-reduction assay, with the MIC of miltefosine defined as the lowest concentration that inhibits 70% of fungal growth.

2.6. Fluorescent Staining to Evaluate Alterations in the Fungal Cell

Calcofluor White (Sigma-Aldrich, St. Louis, MO, USA), concanavalin A (Sigma-Aldrich, St. Louis, MO, USA), Nile red (Sigma-Aldrich, St. Louis, MO, USA), 2′,7′-dichlorofluorescein diacetate (DCFH-DA) (Sigma-Aldrich, St. Louis, MO, USA), and JC-1 (ThermoFisher, Waltham, MA, USA) were used as fluorescent probes to evaluate alterations in cell wall sugars, membrane lipids, ROS production, and mitochondrial membrane potential, respectively. Fungal conidia (1×10^5) were grown for 48 h at 37 °C in 5% CO_2 in the presence of miltefosine (1/2 MIC_{70}), and untreated cells were used as control. After a washing step, cells were stained with these fluorescent probes for 1 h at 37 °C in the dark. Then, the samples were washed three times to remove residual dye and 1×10^5 cells were suspended in PBS. The fluorescence intensity was measured using the SpectraMax 340 microplate reader (Molecular Devices, San Jose, CA, USA) at the following wavelengths:

Calcofluor White at 350 nm (excitation) and 432 nm (emission), concanavalin A at 495 nm (excitation) and 520 nm (emission), Nile Red at 550 nm (excitation) and 550 nm (emission), DCFH-DA at 492 nm (excitation) and 517 nm (emission), JC-1 at 475 nm (excitation) and 529 nm (green fluorescence), and 590 nm (red fluorescence) for the calculation of the red/green fluorescence intensity.

2.7. Transmission Electron Microscopy

The cells were centrifuged, washed with sterile PBS buffer, and processed for TEM via the following steps: (i) fixation in 2.5% glutaraldehyde, 4% formaldehyde, and 0.01 M calcium chloride in 0.1 M cacodylate buffer for 2 h at room temperature; (ii) three washes in 0.1 M cacodylate buffer; (iii) post-fixation with 1% buffered osmium tetroxide and 0.8% potassium ferrocyanide for 2 h at room temperature; (iv) three washes in 0.1 M cacodylate buffer; (v) sequential dehydration in acetone at 30%, 50%, 70% + 2.5%, 90%, and 100% uranyl acetate 3×; and (vi) embedded in Spurr resin. Ultrathin sections were prepared using an EM UC6 microtome (Leica Microsystems, Wetzlar, Germany), recovered on 300-mesh copper grids, stained with uranyl acetate and lead citrate, and observed on an FEI Morgagni transmission electron microscope (FEI Company, Hillsboro, OR, USA) operated at 80 kV.

2.8. Scanning Electron Microscopy

Mucorales species were grown in RPMI in the presence of 1/2 MIC_{70} of miltefosine with orbital agitation (150 rpm) for 72 h. Positive control consisted of cells grown in the absence of miltefosine. Mycelium samples were gently collected and processed via the following steps: (i) fixation in 2.5% glutaraldehyde and 4% formaldehyde in 0.1 M cacodylate buffer for at least 24 h at 4 °C, (ii) post-fixation in 1% osmium tetroxide in 0.1 M cacodylate buffer containing 1.25% potassium ferrocyanide and 5 mM $CaCl_2$ for 30 min, (iii) dehydration in a graded ethanol series (30–100%), (iv) critical point drying in CO_2, (v) adhesion of samples on aluminum stubs with a carbon tape, and (vi) coating with gold. Images were obtained in the scanning electron microscopes FEI Quanta 250 (FEI Company, Hillsboro, OR, USA) and Zeiss EVO 10 (Zeiss Company, Oberkochen, Germany). Images were processed using Photoshop software (version 24.6.0, Adobe, San Jose, CA, USA).

2.9. Antifungal Drug Synergy Assay

Synergistic interactions were detected by the checkerboard method according to EUCAST guidelines [25]. Conidia of Mucorales species (3×10^4) were grown in 96-well plates containing RPMI in the presence of miltefosine (0.125–8 μg/mL) combined with posaconazole (0.5–64 μg/mL) or amphotericin B (0.78–50 μg/mL). After 72 h incubation at 37 °C in 5% CO_2, the MIC_{70} was determined by readings at 600 nm and cell viability was assessed by XTT-reduction assay at 490 nm. Interactions were determined by two different methods: the fractional inhibitory concentration index (FICI) and the Bliss independence model.

The fractional inhibitory concentration index was calculated using the following formula: (MIC_{70} combined/MIC_{70} drug A alone) + (MIC_{70} combined/MIC_{70} drug B alone). The results were classified as synergistic effect, FICI of $\leq$0.5; no effect, FICI of >0.5–4.0; and antagonistic effect, FICI of >4.0 [28].

The Bliss independence model was performed according to Meletiadis and colleagues and Zhao and colleagues [29,30]. The following formula was used to assess the drug interaction: $Eexp = Ea + Eb - Ea \times Eb$, in which Eexp is the expected efficacy of drug combination, Ea is the efficacy of drug A (miltefosine), and Eb is the efficacy of drug B (amphotericin B and posaconazole). The results were classified as synergistic effect, Eobs > Eexp; indifference, Eobs = Eexp; and antagonistic effect, Eobs < Eexp.

2.10. Statistical Analyses

All experiments were performed in triplicate in three independent experimental sets. The experimental results are presented as the mean ± standard deviation (SD). Data were analyzed by nonparametric Kruskal–Wallis one-way analysis of variance to compare the differences among the groups (the group treated with miltefosine and the control group without miltefosine). The individual comparisons of the groups were performed using a Bonferroni post-test. Statistical analyses were performed using GraphPad Prism v5.00 for Windows (GraphPad Software, San Diego, CA, USA). The 90% or 95% confidence interval was determined in all experiments, considering $p < 0.05$ a statistically significant difference.

3. Results

3.1. Susceptibility of Mucorales Species to Miltefosine

The minimum inhibitory concentration of miltefosine varied from 2 µg/mL for *M. velutinosus* and 4 µg/mL for the three *Rhizopus* species to 8 µg/mL for *Cunninghamella* spp. (Table 1). The Mucorales species tested were more susceptible to posaconazole than amphotericin B, which were used as a control since these antifungal drugs are usually chosen in clinical settings for the treatment of Mucorales infections (Table 1). Although MIC values for posaconazole were lower than the MIC for amphotericin B, only amphotericin B showed a fungicidal effect against the tree *Rhizopus* species at 25 µg/mL (Table 1). *Cunninghamella* spp. was the species most resistant to miltefosine and amphotericin B but was susceptible to posaconazole. *M. velutinosus* was susceptible to miltefosine at 1 µg/mL, the lowest MIC values for miltefosine found in this study, and susceptible to posaconazole at 2 µg/mL, but it was resistant to amphotericin B at 25 µg/mL (Table 1). Despite being susceptible to miltefosine and posaconazole, all drugs tested showed a fungistatic effect for *M. velutinosus*. The analysis of minimum fungicidal concentration revealed that all species remained viable in the presence of miltefosine, suggesting that it demonstrates fungistatic activity (Table 1). However, even posaconazole, which is used in clinical settings, displayed only a fungistatic effect against the Mucorales samples used in the study (Table 1).

Table 1. Minimum inhibitory and fungicidal concentration of miltefosine, amphotericin B, and posaconazole against *R. oryzae* UCP1295, *R. stolonifer* UCP1300, *R. microspores* UCP1304, *Cunninghamella* sp. B926, and *M. velutinosus* H136BO.

Species	Miltefosine		Amphotericin B		Posaconazole	
	MIC_{70}	MFC	MIC_{70}	MFC	MIC_{70}	MFC
R. oryzae	4 µg/mL	>64 µg/mL	25 µg/mL	25 µg/mL	8 µg/mL	>16 µg/mL
R. stolonifer	4 µg/mL	>64 µg/mL	25 µg/mL	25 µg/mL	8 µg/mL	>16 µg/mL
R. microsporus	4 µg/mL	>64 µg/mL	25 µg/mL	25 µg/mL	4 µg/mL	>16 µg/mL
Cunninghamella spp.	8 µg/mL	>64 µg/mL	>50 µg/mL	>50 µg/mL	1 µg/mL	>16 µg/mL
M. velutinosus	2 µg/mL	>64 µg/mL	25 µg/mL	>50 µg/mL	1 µg/mL	>16 µg/mL

MIC: minimal inhibitory concentration. MFC: minimum fungicidal concentration.

3.2. Effect of Miltefosine on Biofilms

The potential of miltefosine against fungal biofilms was also evaluated. Regarding the fungal biomass, miltefosine decreased at least 50% of biofilm growth at 4 MIC for *R. oryzae* and *R. stolonifer* and at 8 MIC for *R. microsporus*, *Cunninghamella* spp., and *M. velutinosus* (Figure 1A). Biofilm matrix was 50% reduced at 4 MIC for *R. stolonifer* and at 8 MIC for *R. oryzae*, *R. microsporus*, *Cunninghamella* spp., and *M. velutinosus* (Figure 1B). Miltefosine decreased at least 50% of biofilm viability at 4 MIC for *R. stolonifer*, *Cunninghamella* spp., and *M. velutinosus* and at 8 MIC for *R. oryzae* and *R. microsporus* (Figure 1C). These results suggest that miltefosine not only acts against planktonic cells but also displays antifungal activity against biofilms formed by Mucorales species.

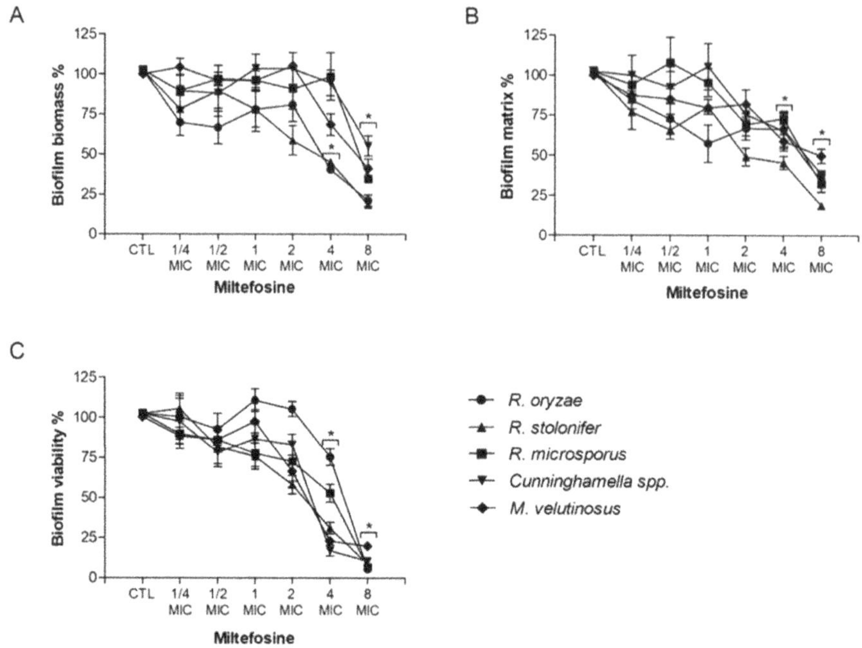

Figure 1. Effect of miltefosine on preformed biofilms of *R. oryzae* UCP1295, *R. stolonifer* UCP1300, *R. microsporus* UCP1304, *Cunninghamella* spp. B926, and *M. velutinosus* H136BO. Fungal biofilm was firstly formed in RPMI medium on polystyrene surface for 24 h and was then treated with different concentrations of miltefosine (1/4–8 MIC) for another 24 h incubation. Intact fungal biofilms were considered controls (CTL, 100% biofilm) and their degradation due to treatment was compared to the control. Fungal biomass (**A**), extracellular matrix (**B**), and viability (**C**) were measured using violet crystal, safranin, and XTT-reduction assay, respectively. * $p < 0.05$, compared to zero (absence of drug) for each species.

3.3. Susceptibility of Miltefosine-Treated Cells to Membrane Stressors

In order to elucidate the modifications that miltefosine causes to the fungal cell surface, the susceptibility to membrane stressors such as SDS and NaCl was evaluated in the presence of miltefosine, and all compounds were used at sub-inhibitory concentrations (Figure 2). Except for *R. oryzae*, which showed no alteration in susceptibility, all species were more susceptible to SDS in the presence of miltefosine (Figure 2A). *R. oryzae*, *Cunninghamella* spp., and *M. velutinosus* were more susceptible to NaCl in the presence of miltefosine (Figure 2B). However, the susceptibility to NaCl was not altered for *R. stolonifer* and *R. microsporus* in the presence of miltefosine (Figure 2B).

3.4. Susceptibility to Miltefosine in the Presence of Exogenous GlcCer and Ergosterol

To check whether miltefosine MIC_{70} values are altered in the presence of exogenous GlcCer and ergosterol, these molecules were added to the media and an antifungal susceptibility test was performed to compare MIC_{70} values of miltefosine alone and in the presence of GlcCer and ergosterol.

In the presence of GlcCer, MIC_{70} for miltefosine increased fourfold for *R. oryzae*, eightfold for *R. microsporus*, fourfold for *R. stolonifer*, eightfold for *Cunninghamella* spp., and twofold for *M. velutinosus* (Table 2). Regarding the addition of ergosterol, MIC_{70} for miltefosine increased at least fourfold for all species when ergosterol was added (Table 2).

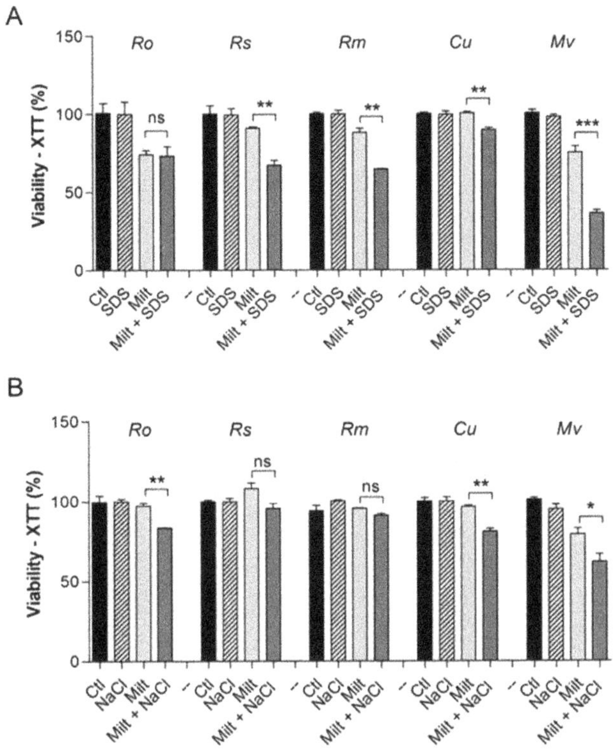

Figure 2. Susceptibility of Mucorales species to Miltefosine in the presence of surface stressors. SDS (90 μg/mL for *Cunninghamella* spp. and 45 μg/mL for the other species) was used as a membrane stressor (**A**). NaCl 1% was used as an osmotic stressor (**B**). The control represents fungal viability in the absence of stressors and miltefosine. Ro: *R. oryzae;* Rs: *R. stolonifer;* Rm: *R. microspores;* Cu: *Cunninghamella* spp.; Mv: *M. velutinosus;* Ctl: control; Milt: miltefosine; SDS: sodium dodecyl sulphate. * $p < 0.05$, ** $p < 0.01$, *** $p < 0.001$. ns—not significant. Experiments were performed in duplicate in three independent experimental sets.

Table 2. The minimum inhibitory concentration of miltefosine when *R. oryzae* UCP1295, *R. stolonifer* UCP1300, *R. microspores* UCP1304, *Cunninghamella* sp. B926, and *M. velutinosus* H136BO were grown in the presence of exogenous GlcCer or ergosterol.

Species	MIC				
	Miltefosine	Miltefosine + GlcCer	Miltefosine + GlcCer	Miltefosine + Ergosterol	Miltefosine + Ergosterol
		50 μg/mL	100 μg/mL	50 μg/mL	100 μg/mL
R. oryzae	4 μg/mL	16 μg/mL	16 μg/mL	16 μg/mL	32 μg/mL
R. stolonifer	4 μg/mL	16 μg/mL	16 μg/mL	16 μg/mL	32 μg/mL
R. microsporus	4 μg/mL	16 μg/mL	16 μg/mL	16 μg/mL	32 μg/mL
Cunninghamella spp.	8 μg/mL	32 μg/mL	32 μg/mL	32 μg/mL	32 μg/mL
M. velutinosus	2 μg/mL	2 μg/mL	4 μg/mL	8 μg/mL	16 μg/mL

MIC: minimal inhibitory concentration. GlcCer: glucosylceramide purified from *R. stolonifer* UCP1300.

3.5. Cell Alterations Caused by Miltefosine

In order to characterize the alterations in Mucorales cells caused by miltefosine, different fluorescent staining procedures were performed. Calcofluor white and concanavalin A were used as cell wall markers due to their ability to evaluate modifications in chitin and mannose residues, respectively. Treatment with miltefosine resulted in a reduction in chitin content for *Cunninghamella* spp. and *M. velutinosus* (Figure 3A), whereas a decrease in mannose residues was observed for all five fungi (Figure 3B).

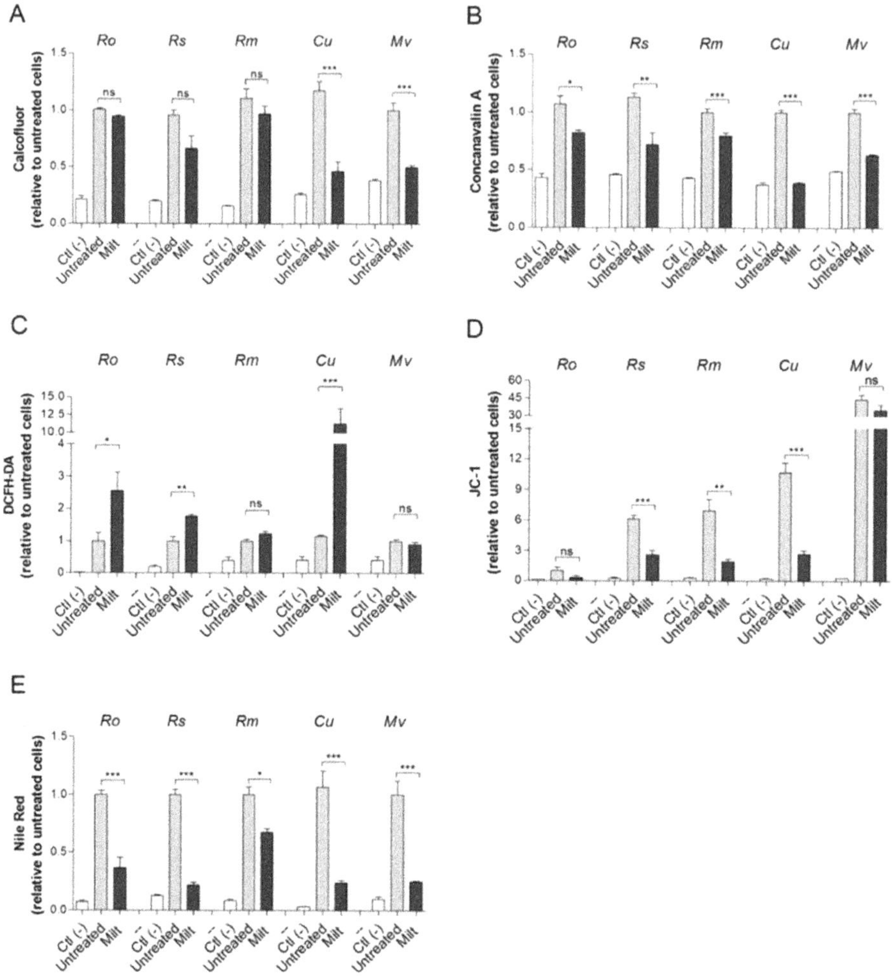

Figure 3. The effect of miltefosine on Mucorales species analyzed by fluorescent staining. Cells were grown in the presence of 1/2 MIC_{70} for 72 h at 37 °C. Chitin content was analyzed using calcofluor white (**A**). Concanavalin A was used to evaluate mannose residues (**B**). Oxidative stress; ROS was measured using DCFH-DA (**C**). The mitochondrial membrane polarization was measured using JC-1 (**D**). Neutral lipids were quantified using Nile Red stain (**E**). Ctl (–), a negative control that represents cells in the absence of fluorescent stain. Untreated, a positive control that represents cells stained with fluorescent stain but without drug treatment. Ro: *R. oryzae*; Rs: *R. stolonifer*; Rm: *R. microspores*; Cu: *Cunninghamella* spp.; Mv: *M. velutinosus*; Milt: miltefosine; ns: not significant. * $p < 0.05$; ** $p < 0.01$; *** $p < 0.001$. Experiments were performed in quadruplicate in three independent experimental sets.

Oxidative stress and depolarization of mitochondrial membranes were also evaluated using DCFH-DA and JC-1 staining, respectively. Miltefosine caused oxidative stress in *R. oryzae*, *R. stolonifer*, and *Cunninghamella* spp. since ROS staining increased compared to the untreated control (Figure 3C). Regarding the mitochondrial membrane, miltefosine led to a decreased depolarization in *R. microsporus*, *R. stolonifer*, and *Cunninghamella* spp. because the JC-1 ratio was reduced compared to untreated control (Figure 3D). Miltefosine treatment reduced the neutral lipid content for all Mucorales species tested since Nile Red staining intensity decreased compared to the untreated control (Figure 3E).

3.6. Transmission Electron Microscopy of Fungal Cells Treated with Miltefosine

Transmission electron microscopy (TEM) was performed to analyze whether morphological and ultrastructural changes are observed when fungal cells are treated with miltefosine 1/2 MIC_{70} values (Figure 4). Mucorales species were grown for 72 h in the absence of miltefosine and exhibited granules with different electron densities and some vacuoles with different sizes (Figure 4A,E,I,M,Q). In conditions treated for 72 h, cytoplasmic extravasation was observed in all cases (Figure 4B–D,F–H,J–L,N–P,R–T).

3.7. Scanning Electron Microscopy of Fungal Cells Treated with Miltefosine

Scanning electron microscopy (SEM) was performed to analyze whether miltefosine induced morphological alterations even at subinhibitory concentrations (miltefosine 1/2 MIC_{70} values) (Figure 5). Mucorales species growing for 72 h in the absence of miltefosine showed a mycelium containing non-septate hyphae, with spores germinated into hyphae (Figure 5A,C,E,G,I). When cells were treated with miltefosine, it was possible to observe an increase in the presence of rupture (Figure 5B) and amorphous cells (arrows in Figure 5D,F,H). The main alteration observed in *M. velutinosus* was the presence of more spores with a small germ tube and fewer developed hyphae (Figure 5H).

3.8. Evaluation of Synergism with Antifungal Drugs

The interaction of miltefosine with posaconazole and amphotericin B was evaluated. According to the FICI, miltefosine presented no effect with posaconazole, which was observed for all five species tested (Table 3). Although FICI did not indicate any positive interaction, a reduction in MIC_{70} values was seen for miltefosine and posaconazole in *R. stolonifer* and *Cunninghamella* spp. (Table 3). Regarding the interaction with amphotericin B, miltefosine showed a synergic effect only for *Cunninghamella* spp. (Table 3).

Table 3. Antifungal activity of miltefosine, posaconazole, and amphotericin B—alone and in combination according to the Fractional Inhibitory Concentration Index (FICI)—against *R. oryzae* UCP1295, *R. stolonifer* UCP1300, *R. microspores* UCP1304, *Cunninghamella* sp. B926, and *M. velutinosus* H136BO. MIC values were used to analyze the interaction between miltefosine with posaconazole or amphotericin B.

Species	MIC_{70} Alone (μg/mL)			MIC_{70} Combined		FICI	
	Milt	Ampho B	Posa	Ampho B/Milt	Posa/Milt	Ampho B + Milt	Posa + Milt
R. oryzae	4	25	8	12.5/0.25	8/4	0.56 (N)	2.0 (N)
R. stolonifer	4	25	8	25/4	2/2	2.0 (N)	0.75 (N)
R. microsporus	4	25	4	12.5/4	4/4	1.5 (N)	2.0 (N)
Cunninghamella spp.	8	100	1	2.5/0.25	0.25/4	0.056 (S)	0.75 (N)
M. velutinosus	2	25	1	6.25/1	1/2	0.75 (N)	2.0 (N)

Milt: miltefosine; Ampho B: amphotericin B; Posa: posaconazole; MIC: minimal inhibitory concentration; S: synergistic interaction; A: antagonist interaction; N: no effect.

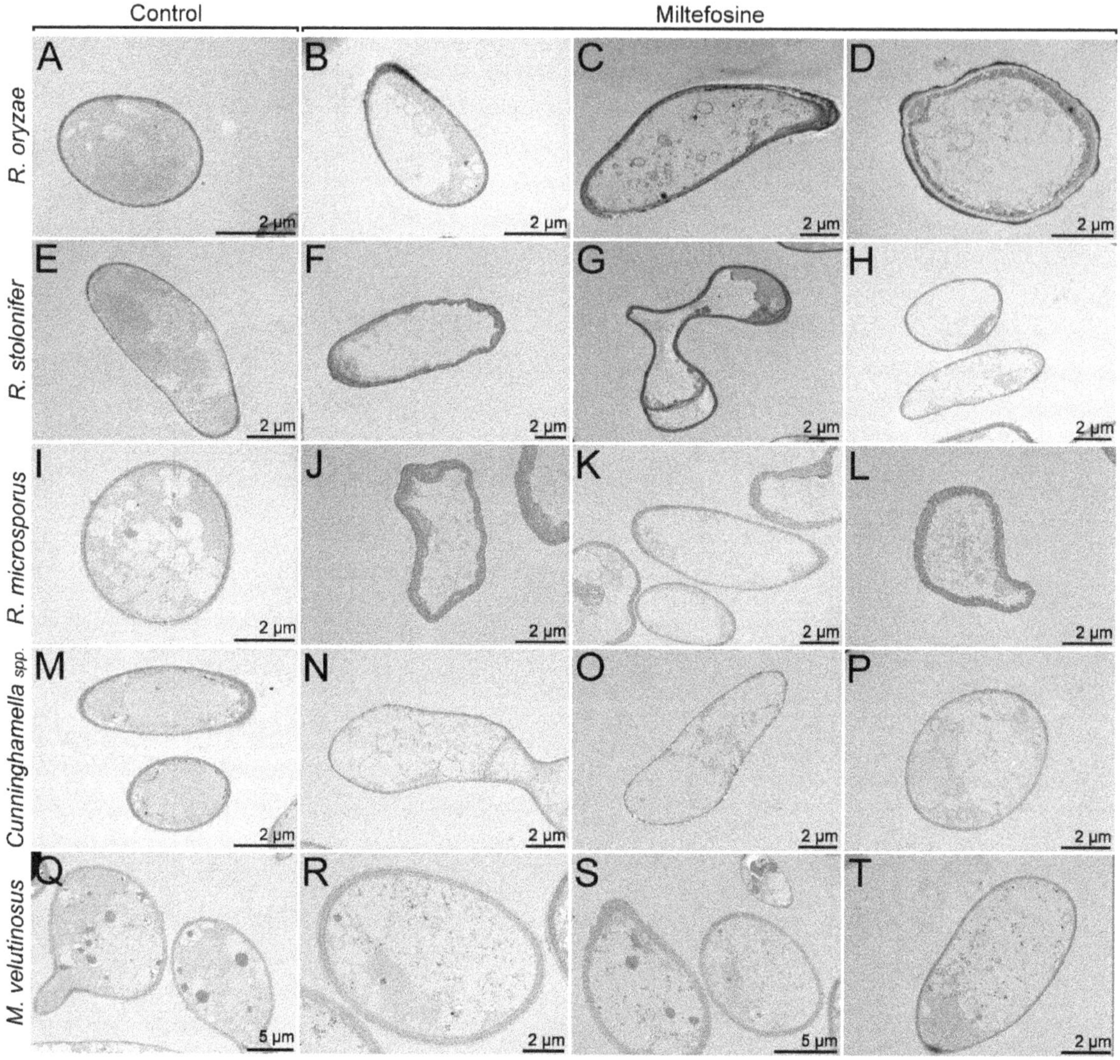

Figure 4. Mucorales alterations after exposure to miltefosine, evaluated by transmission electron microscopy. The control cultures (**A**,**E**,**I**,**M**,**Q**) showed granules with different electron densities and some vacuoles with different sizes. The treated samples (**B**–**D**,**F**–**H**,**J**–**L**,**N**–**P**,**R**–**T**) presented cytoplasmic extravasation.

When the interaction was analyzed by the Bliss method, which considers the percentage of efficacy, miltefosine presented synergistic interaction with posaconazole for *R. oryzae*, *R. stolonifer*, *Cunninghamella* spp., and *M. velutinosus* and antagonistic interaction for *R. microsporus* (Table 4). Miltefosine showed synergistic interaction with amphotericin B for all species tested except for *M. velutinosus*, which presented an indeterminate effect (Table 4).

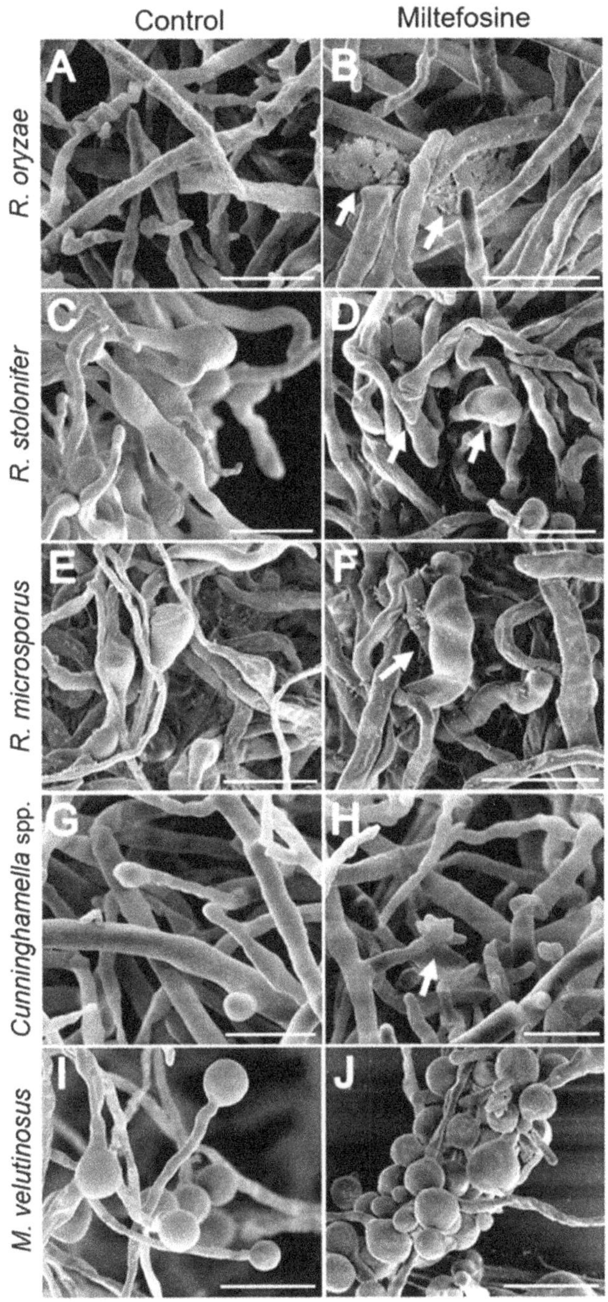

Figure 5. Mucorales alterations after exposure to miltefosine, evaluated by scanning electron microscopy. The control cultures (**A**,**C**,**E**,**G**,**I**) exhibited non-septate hyphae and treated samples (**B**,**D**,**F**,**H**,**J**) showed rupture and amorphous cells (arrows). Bar: 20 μm.

Table 4. Antifungal activity of miltefosine, amphotericin B, and posaconazole—alone and in combination according to the Bliss independence method—against *R. oryzae* UCP1295, *R. microspores* UCP1304, *R. stolonifer* UCP1300, *Cunninghamella* sp. B926, and *M. velutinosus* H136BO.

Species	Efficacy of Drugs Alone (% of Inhibition—½ MIC_{70})			Efficacy of Combined Drugs					
				Ampho B			Posa		
	Milt	Ampho B	Posa	E_{obs}	E_{exp}	ΔE, %	E_{obs}	E_{exp}	ΔE, %
R. oryzae	29.9	45.1	43.3	72.5	61.5	11.0 (S)	91.2	46;4	44.8 (S)
R. microsporus	2.7	35.7	57.3	19.5	18.2	1.3 (S)	20.4	33.5	−13.1 (A)
R. stolonifer	44.3	41.7	64.2	81.2	67.6	13.6 (S)	93.7	71.6	22.1 (S)
Cunninghamella spp.	0.5	0.4	10.3	89.9	1.5	88.4 (S)	82.1	12.1	70.0 (S)
M. velutinosus	38.6	44.8	36	66.1	66.1	0 (I)	51.6	49.3	2.3 (S)

Milt: miltefosine; Ampho B: amphotericin B; Posa: posaconazole; MIC: minimal inhibitory concentration; I: indeterminate; S: synergistic interaction; A: antagonist interaction; E_{obs}: efficacy observed in the analysis; E_{exp}: efficacy expected according to Bliss calculation; ΔE: difference between E_{obs} and E_{exp} (interaction).

4. Discussion

Mucormycosis is a life-threatening infection that is difficult to treat since the causing agents affect mainly immunocompromised patients and are known to be intrinsically resistant to several antifungal drugs, including most azoles [31]. In this context, new options for mucormycosis treatment are needed, and drug repurposing has emerged as an interesting approach.

Miltefosine is a drug approved by the US Food and Drug Administration with an established antiparasitic activity, as it is used to treat visceral and cutaneous leishmaniasis, especially in countries such as India, Colombia, and Brazil [32]. In the last few years, several studies have reported the in vitro efficacy of miltefosine against fungal species, including yeasts and filamentous and dimorphic fungi, indicating broad-spectrum antifungal activity [19,22]. In this work, we describe the antifungal activity of miltefosine against mucormycosis-causing agents, a group of non-septate fungi all belonging to the Mucorales order. Our results show that MIC_{70} values for Mucorales species ranged from 2 to 8 μg/mL, corroborating a previous report in *Rhizopus* spp. that showed MICs from 2 to 16 μg/mL [22]. Unlike observations made with several other fungi, such as *Scedosporium* spp., *Sporothrix* spp., *Cryptococcus* spp., *Paracoccidioides* spp., *C. albicans*, and *F. oxysporum*, miltefosine did not exhibit fungicidal activity in concentrations up to 64 μg/mL against the Mucorales species used in this work [14,18,23,24,33]. This indicates a different manner of antifungal activity against non-septate fungi.

Castro and colleagues (2017) conducted an open-label clinical trial to assess the pharmacokinetic activity of miltefosine (Impavido) at a nominal dose of 2.5 mg/kg/day for 28 days for the oral treatment of cutaneous leishmaniasis [34]. According to the study, the median maximum concentration (Cmax) of miltefosine in plasma observed in children (n = 30) and adults (n = 29) was 22.7 μg/mL and 31.9 μg/mL, respectively [34]. The in vitro MIC_{70} values of miltefosine reported in our study were lower than the Cmax observed by Castro and colleagues in 2017 [34]. If we simply compare concentrations, the dosing regimen for the treatment of cutaneous leishmaniasis may result in a Cmax that would be sufficient for antifungal activity of miltefosine against *R. oryzae*, *R. microsporus*, *R. stolonifer*, *Cunninghamella* spp., and *M. velutinosus*, since Cmax > MIC_{70}. However, a more refined pharmacokinetic/pharmacodynamic (PK/PD) relationship is necessary to better predict the efficacy of an antifungal drug and, consequently, enable the definition of the dosing regimen. Antifungal drugs can exhibit either concentration-dependent or time-dependent activity [35]. This dose–response relationship can be analyzed through PK/PD indexes, such as the Cmax in relation to the MIC (Cmax/MIC; a concentration-dependent measure), the area under the drug concentration curve in relation to MIC (AUC/MIC; a combination of both concentration and time measure), or the fraction of the interval in which the free drug concentration is above the MIC (fT > MIC; a time-dependent measure) [35–37]. As an example, we can mention liposomal amphotericin B, for which the Cmax/MIC is reported as the best PK/PD index to predict clinical response, and it may be ≥4.5 [38]. To identify the best PK/PD index that can predict the efficacy of miltefosine for the treatment of mucormycosis and, thus, design therapeutic dosing regimen, in vivo studies in infections models, for example, will be necessary.

Fungal biofilms are related to increased antifungal resistance and immune system evasion. In this context, mature biofilms are difficult to eradicate, and their presence on different surfaces may lead to new forms of contamination and worse prognosis during infection [39]. Miltefosine also exhibited anti-biofilm activity against Mucorales species, reducing biomass, extracellular matrix, and the viability of preformed biofilms at 4 or 8 MICs, which represents concentrations varying from 8 to 64 μg/mL. These results further establish the anti-biofilm activity of miltefosine since this drug is effective against biofilms of *C. albicans*, *F. oxysporum*, *Scedosporium* spp., and *Sporothrix* spp. [20,23,33,40].

As an alkylphospholipid, miltefosine acts by interacting with membrane lipids and altering membrane functions, as observed in studies using mimetic membranes and parasites. These studies observed that lipid rafts and sterol content play a major role in the interaction of miltefosine with plasma membranes [41,42]. Fungal lipid rafts are composed mainly of ergosterol and sphingolipids, which are important structures for cell division and morphogenesis [43]. In the present work, the addition of exogenous ergosterol or GlcCer (a major fungal sphingolipid) was able to significantly increase the MIC values of miltefosine in all Mucorales species. Previous works showed that exogenous ergosterol also increased miltefosine MIC in *Cryptococcus* spp. and *Candida krusei*, whereas GlcCer increased MIC in *Scedosporium aurantiacum* [23,24,44]. Subinhibitory concentrations of miltefosine were able to increase Mucorales species susceptibility to either SDS or NaCl (used as membrane stressors), indicating that miltefosine impacts the membrane integrity of these species, as has been reported for several other fungi [14,23,24]. These results highlight that part of miltefosine's antifungal properties are related to an interaction with important fungal lipids that leads to an alteration in membrane integrity.

Cell alterations caused by miltefosine treatment were analyzed using fluorescent staining for different parameters. Neutral lipid content was reduced in all Mucorales species. It is known that miltefosine is able to interfere with lipid metabolism in parasites, as previous studies demonstrated that phosphatidylcholine concentration was decreased in *Trypanosoma cruzi* and *Leishmania donovani* [45,46]. Brilhante and colleagues demonstrated that ergosterol concentration was reduced after miltefosine treatment in *Histoplasma capsulatum* and *Coccidioides posadasii*, and a reduction in neutral lipid content was also observed in our previous work with *Scedosporium aurantiacum* [15,23].

Regarding cell wall parameters, chitin content was reduced only in *Cunninghamella* spp. and in *M. velutinosus*, whereas mannose residues were reduced in all five species. Taken together, these results suggest that miltefosine alters important fungal cell wall components. As observed in previous studies, *Cryptococcus* and *Sporothrix* yeasts had thinner cell walls after miltefosine treatment [14,24].

Oxidative stress and mitochondrial alterations have been described as additional mechanisms of action of miltefosine and are related to the induction of apoptosis, as demonstrated in other fungal species. In *Saccharomyces cerevisiae*, miltefosine induces cell death by interacting with COX9 (cytochrome c oxidase complex subunit of the electron transport chain), leading to a decrease in mitochondrial membrane potential [47]. Miltefosine also reduced mitochondrial membrane potential and increased intracellular ROS levels in *C. albicans*, *C. neoformans*, and *S. aurantiacum* [23,24,48]. In our study, the increase in ROS levels and decrease in mitochondrial membrane potential seemed to be species specific, with *Cunninghamella* spp. and *R. stolonifer* having both parameters altered, whereas only *M. velutinosus* had no alteration. Interestingly, even with these alterations being hallmarks of apoptosis and being observed in the Mucorales species used in this work, miltefosine had no fungicidal activity, as mentioned earlier. Xu and colleagues described that *R. oryzae* upregulated genes encoding heat shock proteins and oxidative stress response enzymes (superoxide dismutase and catalase) when intracellular ROS levels were increased [49]. The observations in our study led to hypotheses relating to possible stress response mechanisms that are activated in order to halt cell death processes induced by miltefosine.

The main observation in scanning electron microscopy (SEM) for all species treated with miltefosine was the presence of ruptured and amorphous cells, with the exception of *M. velutinosus*, which presented more spores with a small germ tube and fewer developed hyphae after miltefosine treatment. Previous reports have demonstrated some morphological alterations caused by classical antifungals in yeast and filamentous fungi. Dunyach and colleagues (2011) showed that the treatment of *Candida albicans* with caspofungin led to severe alterations in the cell wall and a loss of cell volume and cytoplasmic content, and consequently, there was no yeast–hypha transition [50]. Morphological alterations were also observed for *Sporothrix brasiliensis* treated with ketoconazole alone or in a complex with zinc, which included yeast–hyphae conversion, an increase in cell size, and cell wall damage to amorphous cells [51]. *Sporothrix schenckii* treated with amphotericin B led to an increase in single yeast and the appearance of amorphous cells compared to the control without treatment [52]. Few reports have demonstrated the morphological alterations in Mucorales species treated with classical antifungal drugs such as amphotericin B and posaconazole. Macedo and colleagues (2019) demonstrated that the combination of voriconazole with amphotericin B, posaconazole, or caspofungin led to small, rounded, and compact hyphal forms of *Rhizopus microsporus* compared to the growth control [53]. A recent study showed the morphological changes caused by selected compounds from the Pandemic Response Box® library from Medicines for Malaria Venture (MMV), which mainly affected sporangia and spore formation from *Rhizopus oryzae* [54].

Drug combination is an important strategy to improve treatment and avoid the emergence of antifungal resistance. Due to the lack of clinical studies, combination therapy is not currently recommended as a first-choice treatment of mucormycosis [55]. Drogari-Apiranthitou and colleagues reported that double-drug combinations of amphotericin B, anidulafungin, and posaconazole were not synergic against *Rhizopus* spp. and *Lichthemia ramosa*. The only exception was the combination of amphotericin B with anidulafungin against *M. circinelloides* [56]. In the present work, we analyzed the efficacy of miltefosine combined with either amphotericin B or posaconazole via two distinct methods. FICI analysis revealed that the combination of miltefosine with amphotericin B was synergic against *Cunninghamella* spp., whereas it had no effect on the other species. However, it is important to note that, when combined with miltefosine, amphotericin B MIC values were reduced in all species except in *R. stolonifer*. The combination of miltefosine and posaconazole was also indifferent for all five species according to FICI. Using the Bliss independence method, miltefosine combined with amphotericin B was shown to act synergistically against all species except for *M. velutinosus*, for which the effect was indifferent. The combination of miltefosine and posaconazole had a synergic effect on all species with the exception of *R. microsporus*, for which the effect was antagonistic. The determination of a synergic effect between two drugs is highly dependent on several aspects, including the species tested and the method of analysis. In vitro, in vivo, and clinical trials might result in contrasting data. For example, the combination of the iron chelator deferasirox with liposomal amphotericin B or micafungin resulted in synergic effects in a diabetic murine model of mucormycosis; however, higher mortality was observed in the combination therapy group during a clinical trial [57,58]. Considering that the Bliss method revealed synergic interactions and that by using the FICI method we observed reductions in amphotericin B MIC values, in vivo studies are very important to see whether these drug combinations would improve the outcomes in infection models. If successful, combined drug therapy would be important for using lower drug concentrations, for preventing adverse effects in patients, and for the emergence of antifungal resistance, as it has already been reported that *Mucor circinelloides* has genes for efflux pumps responsible for azole resistance [59]. Although miltefosine is a drug that induces several cellular alterations and possibly has multiple molecular targets, further studies are needed to clarify the efficacy of combination treatment with the drugs of choice for mucormycosis treatment.

In spite of the results observed, this study has some limitations, such as the elucidation of the chemical interactions between miltefosine with ergosterol and glucosylceramide and the ability of miltefosine to alter the lipid composition of treated fungal cells, as observed for MDCK [60,61] and HepG2 cells [62], *Leishmania donovani* [46], and *Trypanosoma cruzi* [45]. The main limitation of the present study is the lack of an in vivo study to demonstrate the antifungal efficacy of miltefosine in Mucorales.

In conclusion, in the present work, we describe that miltefosine has fungistatic and anti-biofilm activity against five different mucormycosis etiologic agents and that this drug interacts with fungal lipids, destabilizing the plasma membrane and affecting the cell wall. Therefore, miltefosine could be considered for drug repurposing to improve the efficacy of mucormycosis treatment (Figure 6).

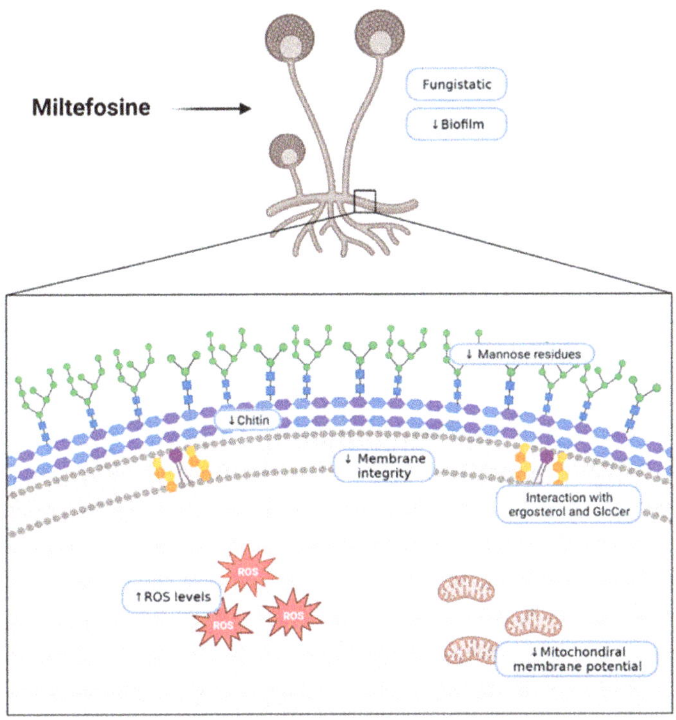

Figure 6. Schematic representation of the effects of miltefosine on fungal cells of the Mucorales species used in this work. The scheme wascreated with BioRender (https://app.biorender.com/illustrations/62b618fecae28690cdc1e912?slideId=b07dc37f-9866-421f-a4e4-3a17cc10ab37, accessed on 3 October 2023).

Author Contributions: Conceptualization, M.I.D.d.S.X. and R.R.-P.; methodology, M.I.D.d.S.X., R.R.-P., V.P.R., Y.d.C.-A., G.M.P.d.S.-F., J.C., F.d.Á.A. and L.P.B.-S.; software, M.I.D.d.S.X., R.R.-P., Y.d.C.-A., V.P.R., G.M.P.d.S.-F. and L.P.B.-S.; validation, M.I.D.d.S.X. and R.R.-P.; formal analysis, M.I.D.d.S.X. and R.R.-P.; investigation, M.I.D.d.S.X., R.R.-P., V.P.R. and Y.d.C.-A.; resources, S.R., F.d.Á.A. and E.B.-B.; data curation, M.I.D.d.S.X., R.R.-P., V.P.R., Y.d.C.-A., J.C., G.M.P.d.S.-F. and L.P.B.-S.; writing—original draft preparation, M.I.D.d.S.X., R.R.-P. and V.P.R.; writing—review and editing, L.P.B.-S., S.R., F.d.Á.A. and E.B.-B.; visualization, M.I.D.d.S.X.; supervision, S.R. and E.B.-B.; project administration E.B.-B.; funding acquisition, S.R. and E.B.-B. All authors have read and agreed to the published version of the manuscript.

Funding: This study was supported by the following Brazilian funding agencies: Coordenação de Aperfeiçoamento de Pessoal de Nível Superior (CAPES, https://www.gov.br/capes/pt-br); Conselho Nacional de Desenvolvimento Científico e Tecnológico (CNPq, https://www.gov.br/cnpq/pt-br), which supported Rochetti, V.P., Castro-Almeida, Y., Cypriano, J., Abreu, F.A., Rozental, S., and Barreto-Bergter, E.); Fundação Carlos Chagas Filho de Amparo à Pesquisa do Estado do Rio de Janeiro (FAPERJ, https://www.faperj.br/) (Rede Micologia, RJ), which supported Xisto, M.I.D.S., Rollin-Pinheiro, R., Santos-Freitas, G.M.P., Borba-Santos, L.P., Rozental, S., and Barreto-Bergter, E. Xisto, M.I.D.S. was also supported by FAPERJ #E-26/202.051/2020; Rollin-Pinheiro, R. was also supported by FAPERJ #E-26/200.591/2022 and #E-26/210.355/2022; and Borba-Santos, L.P. was also supported by FAPERJ #E-26/203.853/2022. The funders had no role in the study design, data collection and analysis, decision to publish, or preparation of the manuscript. Funding support was provided to buy reagents for the development of the study.

Institutional Review Board Statement: Not applicable.

Informed Consent Statement: Not applicable.

Data Availability Statement: Data are contained within the article.

Acknowledgments: The authors thank Walter Oelemann for the English revision.

Conflicts of Interest: The authors declare no conflict of interest.

References

1. Petrikkos, G.; Skiada, A.; Lortholary, O.; Roilides, E.; Walsh, T.J.; Kontoyiannis, D.P. Epidemiology and clinical manifestations of mucormycosis. *Clin. Infect. Dis.* **2012**, *54* (Suppl. S1), S23–S34. [CrossRef] [PubMed]
2. Farmakiotis, D.; Kontoyiannis, D.P. Mucormycoses. *Infect. Dis. Clin. N. Am.* **2016**, *30*, 143–163. [CrossRef] [PubMed]
3. Lewis, R.E.; Kontoyiannis, D.P. Epidemiology and treatment of mucormycosis. *Future Microbiol.* **2013**, *8*, 1163–1175. [CrossRef] [PubMed]
4. Sarvestani, A.S.; Pishdad, G.; Bolandparvaz, S. Predisposing Factors for Mucormycosis in Patients with Diabetes Mellitus; An Experience of 21 Years in Southern Iran. *Bull. Emerg. Trauma* **2013**, *1*, 164–170.
5. Mora-Martínez, A.; Murcia, L.; Rodríguez-Lozano, F.J. Oral Manifestations of Mucormycosis: A Systematic Review. *J. Fungi* **2023**, *9*, 935. [CrossRef] [PubMed]
6. Tissot, F.; Agrawal, S.; Pagano, L.; Petrikkos, G.; Groll, A.H.; Skiada, A.; Lass-Flörl, C.; Calandra, T.; Viscoli, C.; Herbrecht, R. ECIL-6 guidelines for the treatment of invasive candidiasis, aspergillosis and mucormycosis in leukemia and hematopoietic stem cell transplant patients. *Haematologica* **2017**, *102*, 433–444. [CrossRef]
7. Al-Tawfiq, J.A.; Alhumaid, S.; Alshukairi, A.N.; Temsah, M.H.; Barry, M.; Al Mutair, A.; Rabaan, A.A.; Al-Omari, A.; Tirupathi, R.; AlQahtani, M.; et al. COVID-19 and mucormycosis superinfection: The perfect storm. *Infection* **2021**, *49*, 833–853. [CrossRef]
8. Dilek, A.; Ozaras, R.; Ozkaya, S.; Sunbul, M.; Sen, E.I.; Leblebicioglu, H. COVID-19-associated mucormycosis: Case report and systematic review. *Travel. Med. Infect. Dis.* **2021**, *44*, 102148. [CrossRef]
9. Farias, L.; Damasceno, L.S.; Bandeira, S.P.; Barreto, F.K.A.; Leitão, T.; Cavalcanti, L.P.G. COVID-19 associated Mucormycosis (CAM): Should Brazil be on alert? *Rev. Soc. Bras. Med. Trop.* **2021**, *54*, e04102021. [CrossRef]
10. Song, G.; Liang, G.; Liu, W. Fungal Co-infections Associated with Global COVID-19 Pandemic: A Clinical and Diagnostic Perspective from China. *Mycopathologia* **2020**, *185*, 599–606. [CrossRef]
11. Dorlo, T.P.; Balasegaram, M.; Beijnen, J.H.; de Vries, P.J. Miltefosine: A review of its pharmacology and therapeutic efficacy in the treatment of leishmaniasis. *J. Antimicrob. Chemother.* **2012**, *67*, 2576–2597. [CrossRef]
12. Dorlo, T.P.; Huitema, A.D.; Beijnen, J.H.; de Vries, P.J. Optimal dosing of miltefosine in children and adults with visceral leishmaniasis. *Antimicrob. Agents Chemother.* **2012**, *56*, 3864–3872. [CrossRef]
13. Urbina, J.A. Mechanisms of action of lysophospholipid analogues against trypanosomatid parasites. *Trans. R. Soc. Trop. Med. Hyg.* **2006**, *100* (Suppl. S1), S9–S16. [CrossRef]
14. Borba-Santos, L.P.; Gagini, T.; Ishida, K.; de Souza, W.; Rozental, S. Miltefosine is active against Sporothrix brasiliensis isolates with in vitro low susceptibility to amphotericin B or itraconazole. *J. Med. Microbiol.* **2015**, *64*, 415–422. [CrossRef]
15. Brilhante, R.S.; Caetano, E.P.; Lima, R.A.; Castelo Branco, D.S.; Serpa, R.; Oliveira, J.S.; Monteiro, A.J.; Rocha, M.F.; Cordeiro, R.A.; Sidrim, J.J. In vitro antifungal activity of miltefosine and levamisole: Their impact on ergosterol biosynthesis and cell permeability of dimorphic fungi. *J. Appl. Microbiol.* **2015**, *119*, 962–969. [CrossRef]
16. Compain, F.; Botterel, F.; Sitterlé, E.; Paugam, A.; Bougnoux, M.E.; Dannaoui, E. In vitro activity of miltefosine in combination with voriconazole or amphotericin B against clinical isolates of *Scedosporium* spp. *J. Med. Microbiol.* **2015**, *64*, 309–311. [CrossRef]
17. Imbert, S.; Palous, M.; Meyer, I.; Dannaoui, E.; Mazier, D.; Datry, A.; Fekkar, A. In vitro combination of voriconazole and miltefosine against clinically relevant molds. *Antimicrob. Agents Chemother.* **2014**, *58*, 6996–6998. [CrossRef]
18. Rossi, D.C.P.; Spadari, C.C.; Nosanchuk, J.D.; Taborda, C.P.; Ishida, K. Miltefosine is fungicidal to Paracoccidioides spp. yeast cells but subinhibitory concentrations induce melanisation. *Int. J. Antimicrob. Agents* **2017**, *49*, 465–471. [CrossRef]

19. Tong, Z.; Widmer, F.; Sorrell, T.C.; Guse, Z.; Jolliffe, K.A.; Halliday, C.; Lee, O.C.; Kong, F.; Wright, L.C.; Chen, S.C. In vitro activities of miltefosine and two novel antifungal biscationic salts against a panel of 77 dermatophytes. *Antimicrob. Agents Chemother.* **2007**, *51*, 2219–2222. [CrossRef]
20. Vila, T.; Ishida, K.; Seabra, S.H.; Rozental, S. Miltefosine inhibits Candida albicans and non-albicans Candida spp. biofilms and impairs the dispersion of infectious cells. *Int. J. Antimicrob. Agents* **2016**, *48*, 512–520. [CrossRef]
21. Vila, T.V.; Ishida, K.; de Souza, W.; Prousis, K.; Calogeropoulou, T.; Rozental, S. Effect of alkylphospholipids on Candida albicans biofilm formation and maturation. *J. Antimicrob. Chemother.* **2013**, *68*, 113–125. [CrossRef] [PubMed]
22. Widmer, F.; Wright, L.C.; Obando, D.; Handke, R.; Ganendren, R.; Ellis, D.H.; Sorrell, T.C. Hexadecylphosphocholine (miltefosine) has broad-spectrum fungicidal activity and is efficacious in a mouse model of cryptococcosis. *Antimicrob. Agents Chemother.* **2006**, *50*, 414–421. [CrossRef]
23. Rollin-Pinheiro, R.; Almeida, Y.C.; Rochetti, V.P.; Xisto, M.; Borba-Santos, L.P.; Rozental, S.; Barreto-Bergter, E. Miltefosine Against Scedosporium and Lomentospora Species: Antifungal Activity and Its Effects on Fungal Cells. *Front. Cell Infect. Microbiol.* **2021**, *11*, 698662. [CrossRef] [PubMed]
24. Spadari, C.C.; Vila, T.; Rozental, S.; Ishida, K. Miltefosine Has a Postantifungal Effect and Induces Apoptosis in Cryptococcus Yeasts. *Antimicrob. Agents Chemother.* **2018**, *62*, 8. [CrossRef]
25. Taj-Aldeen, S.J.; Salah, H.; Al-Hatmi, A.M.; Hamed, M.; Theelen, B.; van Diepeningen, A.D.; Boekhout, T.; Lass-Flörl, C. In vitro resistance of clinical Fusarium species to amphotericin B and voriconazole using the EUCAST antifungal susceptibility method. *Diagn. Microbiol. Infect. Dis.* **2016**, *85*, 438–443. [CrossRef] [PubMed]
26. Almeida, C.A.; Azevedo, M.M.B.; Chaves, F.C.M.; Roseo de Oliveira, M.; Rodrigues, I.A.; Bizzo, H.R.; Gama, P.E.; Alviano, D.S.; Alviano, C.S. Piper Essential Oils Inhibit Rhizopus oryzae Growth, Biofilm Formation, and Rhizopuspepsin Activity. *Can. J. Infect. Dis. Med. Microbiol.* **2018**, *2018*, 5295619. [CrossRef]
27. Mello, T.P.; Aor, A.C.; Gonçalves, D.S.; Seabra, S.H.; Branquinha, M.H.; Santos, A.L. Assessment of biofilm formation by *Scedosporium apiospermum*, *S. aurantiacum*, *S. minutisporum* and *Lomentospora prolificans*. *Biofouling* **2016**, *32*, 737–749. [CrossRef]
28. Odds, F.C. Synergy, antagonism, and what the chequerboard puts between them. *J. Antimicrob. Chemother.* **2003**, *52*, 1. [CrossRef]
29. Meletiadis, J.; Petraitis, V.; Petraitiene, R.; Lin, P.; Stergiopoulou, T.; Kelaher, A.M.; Sein, T.; Schaufele, R.L.; Bacher, J.; Walsh, T.J. Triazole-polyene antagonism in experimental invasive pulmonary aspergillosis: In vitro and in vivo correlation. *J. Infect. Dis.* **2006**, *194*, 1008–1018. [CrossRef]
30. Zhao, W.; Sachsenmeier, K.; Zhang, L.; Sult, E.; Hollingsworth, R.E.; Yang, H. A New Bliss Independence Model to Analyze Drug Combination Data. *J. Biomol. Screen.* **2014**, *19*, 817–821. [CrossRef]
31. Hassan, M.I.A.; Voigt, K. Pathogenicity patterns of mucormycosis: Epidemiology, interaction with immune cells and virulence factors. *Med. Mycol.* **2019**, *57*, S245–S256. [CrossRef]
32. Ware, J.M.; O'Connell, E.M.; Brown, T.; Wetzler, L.; Talaat, K.R.; Nutman, T.B.; Nash, T.E. Efficacy and Tolerability of Miltefosine in the Treatment of Cutaneous Leishmaniasis. *Clin. Infect. Dis.* **2021**, *73*, e2457–e2562. [CrossRef]
33. Vila, T.V.; Chaturvedi, A.K.; Rozental, S.; Lopez-Ribot, J.L. In Vitro Activity of Miltefosine against Candida albicans under Planktonic and Biofilm Growth Conditions and In Vivo Efficacy in a Murine Model of Oral Candidiasis. *Antimicrob. Agents Chemother.* **2015**, *59*, 7611–7620. [CrossRef] [PubMed]
34. Castro, M.D.; Gomez, M.A.; Kip, A.E.; Cossio, A.; Ortiz, E.; Navas, A.; Dorlo, T.P.; Saravia, N.G. Pharmacokinetics of Miltefosine in Children and Adults with Cutaneous Leishmaniasis. *Antimicrob. Agents Chemother.* **2017**, *61*, 3. [CrossRef] [PubMed]
35. Howard, A.; Hope, W. Assessment of Antifungal Pharmacodynamics. *J. Fungi* **2023**, *9*, 192. [CrossRef]
36. Gonzalez, J.M.; Rodriguez, C.A.; Agudelo, M.; Zuluaga, A.F.; Vesga, O. Antifungal pharmacodynamics: Latin America's perspective. *Braz. J. Infect. Dis.* **2017**, *21*, 79–87. [CrossRef] [PubMed]
37. Lepak, A.J.; Andes, D.R. Antifungal pharmacokinetics and pharmacodynamics. *Cold Spring Harb. Perspect. Med.* **2014**, *5*, a019653. [CrossRef]
38. Gómez-López, A. Antifungal therapeutic drug monitoring: Focus on drugs without a clear recommendation. *Clin. Microbiol. Infect.* **2020**, *26*, 1481–1487. [CrossRef] [PubMed]
39. Borghi, E.; Borgo, F.; Morace, G. Fungal Biofilms: Update on Resistance. *Adv. Exp. Med. Biol.* **2016**, *931*, 37–47. [CrossRef]
40. Dos Santos, G.M.P.; Borba-Santos, L.P.; Vila, T.; Ferreira Gremião, I.D.; Pereira, S.A.; De Souza, W.; Rozental, S. Sporothrix spp. Biofilms Impact in the Zoonotic Transmission Route: Feline Claws Associated Biofilms, Itraconazole Tolerance, and Potential Repurposing for Miltefosine. *Pathogens* **2022**, *11*, 206. [CrossRef]
41. de Sá, M.M.; Sresht, V.; Rangel-Yagui, C.O.; Blankschtein, D. Understanding Miltefosine-Membrane Interactions Using Molecular Dynamics Simulations. *Langmuir* **2015**, *31*, 4503–4512. [CrossRef] [PubMed]
42. Saint-Pierre-Chazalet, M.; Ben Brahim, M.; Le Moyec, L.; Bories, C.; Rakotomanga, M.; Loiseau, P.M. Membrane sterol depletion impairs miltefosine action in wild-type and miltefosine-resistant Leishmania donovani promastigotes. *J. Antimicrob. Chemother.* **2009**, *64*, 993–1001. [CrossRef] [PubMed]
43. Alvarez, F.J.; Douglas, L.M.; Konopka, J.B. Sterol-rich plasma membrane domains in fungi. *Eukaryot. Cell* **2007**, *6*, 755–763. [CrossRef] [PubMed]
44. Wu, Y.; Wu, M.; Gao, J.; Ying, C. Antifungal Activity and Mode of Action of Miltefosine Against Clinical Isolates of Candida krusei. *Front. Microbiol.* **2020**, *11*, 854. [CrossRef] [PubMed]

45. Lira, R.; Contreras, L.M.; Rita, R.M.; Urbina, J.A. Mechanism of action of anti-proliferative lysophospholipid analogues against the protozoan parasite Trypanosoma cruzi: Potentiation of in vitro activity by the sterol biosynthesis inhibitor ketoconazole. *J. Antimicrob. Chemother.* **2001**, *47*, 537–546. [CrossRef] [PubMed]
46. Rakotomanga, M.; Blanc, S.; Gaudin, K.; Chaminade, P.; Loiseau, P.M. Miltefosine affects lipid metabolism in Leishmania donovani promastigotes. *Antimicrob. Agents Chemother.* **2007**, *51*, 1425–1430. [CrossRef]
47. Zuo, X.; Djordjevic, J.T.; Bijosono Oei, J.; Desmarini, D.; Schibeci, S.D.; Jolliffe, K.A.; Sorrell, T.C. Miltefosine induces apoptosis-like cell death in yeast via Cox9p in cytochrome c oxidase. *Mol. Pharmacol.* **2011**, *80*, 476–485. [CrossRef]
48. Wu, Y.; Chen, Y.; Lu, H.; Ying, C. Miltefosine exhibits fungicidal activity through oxidative stress generation and Aif1 activation in Candida albicans. *Int. J. Antimicrob. Agents* **2023**, *62*, 106819. [CrossRef]
49. Xu, Q.; Fu, Y.; Li, S.; Jiang, L.; Rongfeng, G.; Huang, H. Integrated transcriptomic and metabolomic analysis of Rhizopus oryzae with different morphologies. *Process Biochem.* **2018**, *64*, 74–82. [CrossRef]
50. Dunyach, C.; Drakulovski, P.; Bertout, S.; Jouvert, S.; Reynes, J.; Mallié, M. Fungicidal activity and morphological alterations of Candida albicans induced by echinocandins: Study of strains with reduced caspofungin susceptibility. *Mycoses* **2011**, *54*, e62–e68. [CrossRef]
51. de Azevedo-França, J.A.; Borba-Santos, L.P.; de Almeida Pimentel, G.; Franco, C.H.J.; Souza, C.; de Almeida Celestino, J.; de Menezes, E.F.; Dos Santos, N.P.; Vieira, E.G.; Ferreira, A.; et al. Antifungal promising agents of zinc(II) and copper(II) derivatives based on azole drug. *J. Inorg. Biochem.* **2021**, *219*, 111401. [CrossRef]
52. Borba-Santos, L.P.; Nucci, M.; Ferreira-Pereira, A.; Rozental, S. Anti-Sporothrix activity of ibuprofen combined with antifungal. *Braz. J. Microbiol.* **2021**, *52*, 101–106. [CrossRef] [PubMed]
53. Macedo, D.; Leonardelli, F.; Dudiuk, C.; Vitale, R.G.; Del Valle, E.; Giusiano, G.; Gamarra, S.; Garcia-Effron, G. In Vitro and In Vivo Evaluation of Voriconazole-Containing Antifungal Combinations against Mucorales Using a Galleria mellonella Model of Mucormycosis. *J. Fungi* **2019**, *5*, 5. [CrossRef]
54. Xisto, M.; Rollin-Pinheiro, R.; de Castro-Almeida, Y.; Dos Santos-Freitas, G.M.P.; Rochetti, V.P.; Borba-Santos, L.P.; da Silva Fontes, Y.; Ferreira-Pereira, A.; Rozental, S.; Barreto-Bergter, E. Promising Antifungal Molecules against Mucormycosis Agents Identified from Pandemic Response Box(®): In Vitro and In Silico Analyses. *J. Fungi* **2023**, *9*, 187. [CrossRef] [PubMed]
55. Dannaoui, E. Antifungal resistance in mucorales. *Int. J. Antimicrob. Agents* **2017**, *50*, 617–621. [CrossRef] [PubMed]
56. Drogari-Apiranthitou, M.; Mantopoulou, F.D.; Skiada, A.; Kanioura, L.; Grammatikou, M.; Vrioni, G.; Mitroussia-Ziouva, A.; Tsakris, A.; Petrikkos, G. In vitro antifungal susceptibility of filamentous fungi causing rare infections: Synergy testing of amphotericin B, posaconazole and anidulafungin in pairs. *J. Antimicrob. Chemother.* **2012**, *67*, 1937–1940. [CrossRef]
57. Ibrahim, A.S.; Gebremariam, T.; Luo, G.; Fu, Y.; French, S.W.; Edwards, J.E., Jr.; Spellberg, B. Combination therapy of murine mucormycosis or aspergillosis with iron chelation, polyenes, and echinocandins. *Antimicrob. Agents Chemother.* **2011**, *55*, 1768–1770. [CrossRef]
58. Spellberg, B.; Ibrahim, A.S.; Chin-Hong, P.V.; Kontoyiannis, D.P.; Morris, M.I.; Perfect, J.R.; Fredricks, D.; Brass, E.P. The Deferasirox-AmBisome Therapy for Mucormycosis (DEFEAT Mucor) study: A randomized, double-blinded, placebo-controlled trial. *J. Antimicrob. Chemother.* **2012**, *67*, 715–722. [CrossRef]
59. Nagy, G.; Kiss, S.; Varghese, R.; Bauer, K.; Szebenyi, C.; Kocsubé, S.; Homa, M.; Bodai, L.; Zsindely, N.; Nagy, G.; et al. Characterization of Three Pleiotropic Drug Resistance Transporter Genes and Their Participation in the Azole Resistance of Mucor circinelloides. *Front. Cell Infect. Microbiol.* **2021**, *11*, 660347. [CrossRef]
60. Haase, R.; Wieder, T.; Geilen, C.C.; Reutter, W. The phospholipid analogue hexadecylphosphocholine inhibits phosphatidylcholine biosynthesis in Madin-Darby canine kidney cells. *FEBS Lett.* **1991**, *288*, 129–132. [CrossRef]
61. Wieder, T.; Haase, A.; Geilen, C.C.; Orfanos, C.E. The effect of two synthetic phospholipids on cell proliferation and phosphatidylcholine biosynthesis in Madin-Darby canine kidney cells. *Lipids* **1995**, *30*, 389–393. [CrossRef] [PubMed]
62. Jiménez-López, J.M.; Carrasco, M.P.; Segovia, J.L.; Marco, C. Hexadecylphosphocholine inhibits phosphatidylcholine biosynthesis and the proliferation of HepG2 cells. *Eur. J. Biochem.* **2002**, *269*, 4649–4655. [CrossRef] [PubMed]

Journal of *Fungi*

MDPI

Article

Importance of the *Aspergillus fumigatus* Mismatch Repair Protein Msh6 in Antifungal Resistance Development

Jose Lucio [1,†], Irene Gonzalez-Jimenez [1,†], Alejandra Roldan [1], Jorge Amich [1], Laura Alcazar-Fuoli [1,2] and Emilia Mellado [1,2,*]

1 Mycology Reference Laboratory, National Centre for Microbiology, Instituto de Salud Carlos III (ISCIII), 28220 Majadahonda, Spain; joslucio2@gmail.com (J.L.); alex7799roldan@gmail.com (A.R.); jamich@isciii.es (J.A.); lalcazar@isciii.es (L.A.-F.)

2 CIBER de Enfermedades Infecciosas (CIBERINFEC-CB21/13/00105), Instituto de Salud Carlos III (ISCIII), 28029 Madrid, Spain

* Correspondence: emellado@isciii.es

† These authors contributed equally to this work.

Abstract: One of the systems responsible for the recognition and repair of mistakes occurring during cell replication is the DNA mismatch repair (MMR) system. Two major protein complexes constitute the MMR pathway: MutS and MutL. Here, we investigated the possible relation of four *A. fumigatus* MMR genes (*msh*2, *msh*6, *pms*1, and *mlh*1) with the development of azole resistance related to the phenomenon of multi-drug resistance. We examined the MMR gene variations in 163 *Aspergillus fumigatus* genomes. Our analysis showed that genes *msh*2, *pms*1, and *mlh*1 have low genetic variability and do not seem to correlate with drug resistance. In contrast, there is a nonsynonymous mutation (G240A) in the *msh*6 gene that is harbored by 42% of the strains, most of them also harboring the TR_{34}/L98H azole resistance mechanism in *cyp*51A. The *msh*6 gene was deleted in the *aku*B^{KU80} *A. fumigatus* strain, and the Δ*msh*6 isolates were analyzed for fitness, azole susceptibility, and virulence capacity, showing no differences compared with the *aku*B^{KU80} parental strain. Wild-type *msh*6 and Δ*msh*6 strains were grown on high concentrations of azole and other non-azole fungicides used in crop protection. A 10- and 2-fold higher mutation frequency in genes that confer resistance to boscalid and benomyl, respectively, were observed in Δ*msh*6 strains compared to the wild-type. This study suggests a link between Msh6 and fungicide resistance acquisition.

Keywords: *Aspergillus fumigatus*; antifungal resistance; *msh*6; MMR pathway

1. Introduction

Triazoles are the main drugs used for treatment and prophylaxis against *Aspergillus* spp. infections [1,2]. However, the increased worldwide detection of azole-resistant *Aspergillus fumigatus* isolates is threatening azole drugs' effectiveness in aspergillosis management [3]. Triazole drugs belong to the demethylation inhibitors drug family (DMIs), which targets fungi at the 14-α sterol demethylase (Cyp51), an essential enzyme from the ergosterol biosynthesis route [4]. In the clinical setting, the four triazole drugs used as first-line antifungal treatment are mainly voriconazole and isavuconazole, and also itraconazole and posaconazole. On the other hand, in the agricultural environment, DMIs are used to protect crops against different fungal plant pathogens, but due to their chemical structure similar to clinical azoles, cross-resistance between both drug families is common [5–7]. In addition, other classes of fungicides, such as benzimidazoles (MBCs), strobilurins (QoIs), and succinate dehydrogenase inhibitors (SDHIs), are also intensively used to protect crops, together with DMIs, creating an environmental pressure to the evolution of resistance against all major classes of fungicides [8].

The development of azole resistance in *A. fumigatus* occurs due to a selective pressure of antifungals that can be applied in two different scenarios: the clinical path, usually as a result of long periods of exposure to azole clinical therapy, which associates with single point mutations in *cyp*51A, and an environmental acquisition due to the use of DMIs in the agricultural setting, related to a combination of Cyp51A modifications (several *cyp*51A point mutations together with tandem repeat insertions (TR), in the promoter of *cyp*51A) [9]. The majority of the azole-resistant isolates described to date have been associated with the environmental route, whereas the resistant isolates developed in clinics have arisen after long azole treatments [10]. In plant pathogens, DMI resistance shows a wide variety of mechanisms, although the combination of Cyp51 point mutations and promoter insertions is predominant among diverse fungal species [9].

Unlike DMIs, the impact of non-azole pesticides in the development of *A. fumigatus* azole resistance has been poorly studied [8,11,12]. Some of these non-azole fungicides mentioned previously, including QoIs, SDHIs, and MBCs, are widely used [13], and some studies have reported resistant strains harboring point mutations in their target genes. Resistance mechanisms to QoIs have been associated with point mutations in the cytochrome *bc*1 gene (*cyt*B) [14,15]; SDHIs resistance mechanisms have been related to point mutations in the *sdh*B subunit [15,16]; and lastly, point mutations in the β-tubulin gene (*ben*A) have been linked to resistance to MBCs [17]. We have recently used a collection of *A. fumigatus* strains to reveal a correlation between resistance to these fungicides and the azole resistance mechanisms involving TRs, showing a multi-resistance pattern phenomenon in *A. fumigatus* that could be related to their specific genetic background [8].

In previous whole genome sequencing (WGS) phylogenetic analysis performed in our lab, a genome collection of azole-susceptible and resistant *A. fumigatus* strains from diverse geographical origins was bundled into four different clusters. The most remarkable finding was that all genomes harboring the azole resistance mechanisms with TRs in *cyp*51A promoter were grouped in the same cluster [8,18], suggesting genetic closeness possibly due to additional genetic mechanisms operating in these genomes that could include fitter genotypes. Some studies have suggested genetic instability as a possible mechanism of evolving azole resistance in *A. fumigatus* [19,20], as well as in other fungal pathogens [21–23]. In this sense, the avoidance of mistakes during cell replication allows for genetic stability, which is key for the enduring survival of organisms [24]. To ensure genome stability and to sense and repair damaged DNA, cells use a global signaling network named the DNA damage response (DDR), which includes cell cycle checkpoints, chromatin remodeling, DNA repair (MMR), and DNA-damage tolerance pathways (DDT) [19,25]. Activation of DNA damage checkpoint is the core event of DDR, and in *S. cerevisiae*, this activation involves two highly conserved protein kinases, ScMec1 (ATM) and ScTel1 (ATR), that initiate evolutionarily conserved signal transduction cascades, and repair of DNA damage [25]. The role of these kinases in *A. fumigatus* has been previously investigated, suggesting that genetic instability caused by Δ*atm*A and Δ*atr*A mutations can confer an adaptive advantage, mainly in the intensity of voriconazole resistance acquisition [19].

The DNA mismatch repair (MMR) system conducts the recognition and repair of any mistakes occurring during DNA replication [26]. This pathway is constituted by two protein complexes: MutS, which recognizes and binds to the mismatch, and MutL, which interacts with the mistake and removes the strand [27]. Multiple homologs of these proteins have been identified among the eukaryotic microorganisms, best characterized in *Saccharomyces cerevisiae* [28]. Although the MMR has been poorly investigated in *A. fumigatus* compared to *S. cerevisiae*, the genes comprising it are well conserved through all phylum *Ascomycota*, including *msh*1, *msh*2, *msh*3, *msh*4, *msh*6, and *mlh*1 [29]. Defects in MMR, such as gene loss or alterations of the pathways, can cause alterations in the recognition and repair of mutations, creating, in this scenario, an increased mutation rate associated with mutator phenotypes [28–30].

In this study, we investigated four *A. fumigatus* MMR genes, *msh*2, *msh*6, *pms*1, and *mlh*1, to evaluate their possible implication in genetic instability and thus their potential contribution to the development of antifungal resistance by searching for mutations in these genes among a collection of environmental and clinical *A. fumigatus* isolates. In view of the results obtained, we deleted *msh6* to evaluate the importance of genetic instability in *A. fumigatus* as a possible mechanism of evolving antifungal resistance. A comparison of mutation frequencies in *A. fumigatus* wild-type and Δ*msh6* strains showed that the loss of Msh6 results in increased mutation frequencies when exposed to some antifungals, allowing for a rapid generation of resistance.

2. Materials and Methods

2.1. Aspergillus Fumigatus Strains and Culture

A total of 144 clinical *A. fumigatus* strains obtained over a period of 20 years (2001 to 2021) belonging to the Spanish National Center of Microbiology were used in this study. This collection comprises a heterogeneous group of isolates, including azole-susceptible and azole-resistant strains. Resistant strains include a selection of the most common azole resistance mechanisms described to date (G54, M220, G448S, TR_{34}/L98H, TR_{34}/L98H/S297T/F495I, TR_{46}/Y121F/T289A, TR_{53}). Species-level identification was performed by PCR amplification and sequencing of the ITS1-5.8S-ITS2 regions and a portion of the β-tubulin gene [31]. The reference strain *aku*B^{KU80} was used to perform all the transformation experiments.

The *A. fumigatus* strains were subcultured on potato dextrose agar (PDA) (Oxoid, Madrid, Spain) and incubated at 37 °C until the colonies were visually sporulated (48–72 h). For each experiment, conidia were freshly harvested from a PDA medium with 0.1% Tween 20 (Sigma-Aldrich, Madrid, Spain). The concentration of conidia was measured using hemocytometer counting. For long-lasting conservation, *A. fumigatus* spore suspensions were mixed with an equal volume of autoclaved 50% glycerol, and 0.5 mL aliquots of the mix were stored in cryogenic vials at −80 °C.

2.2. Whole Genome Sequencing Analysis

A search for variants or mutations in MMR genes was performed within a collection of 163 *A. fumigatus* genomes that had been previously sequenced in our laboratory using Nextera® XT Library Prep Kit (Illumina Inc., San Diego, CA, USA) as described before or obtained from public databases [8].

The Illumina reads were trimmed using Trimmomatic (version 0.32) [32]. The sequencing adapters and sequences with low-quality scores on 30 ends (Phred score [Q], <20) were trimmed. Raw Illumina WGS reads were quality-checked by performing quality control with FastQC (version 0.11.3; Babraham Institute, Cambridge, UK). Data sets were analyzed against the *A. fumigatus* A1163 reference genome (GenBank accession number ABDB00000000.1) using WGS-Outbreaker v1.0 (Instituto de Salud Carlos III, Madrid, Spain) (https://github.com/BU-ISCIII/WGS-Outbreaker, accessed on 9 October 2018) with default parameters. The pipeline comprised all steps needed for single nucleotide variant (SNV) analysis using whole-genome sequencing data. Mapping against the genome reference was performed with bwa mem (version 0.7.12-r1039) [33], duplicated reads were removed using Picard (version 1.140) (http://broadinstitute.github.io/picard, accessed on 9 October 2018), and the bedtools coverage v2.26 program [34], was used to perform further quality controls. Here, in order to identify genetic variations among strains, SNV detection (variant calling) and SNV matrix generation were performed using GATK version 3.8.0 [35], with best-practices parameters. The ENSEMBL variant effect predictor script (version 88) was used for variant annotation. Part of the whole genome sequencing project has been deposited in NCBI SRA (project accession number SRP151231). All the data used in this analysis, including information about clinical antifungal susceptibility and azole resistance mechanisms of the strains, can be found in Supplementary Table S1.

2.3. Phylogenetic Analysis and Single-Nucleotide Variant Comparison

The final step of the WGS-Outbreaker pipeline comprised Maximum-likelihood tree construction using RaxML software (version 8.2.9) [36], with a GTRCAT model and 100 bootstrap replicates. Phylogenetic trees were visualized and annotated using the ggtree R package [37], and SNV comparisons were performed using a custom R script, mapping all genomes to the A1163 reference genome.

2.4. Generation of A. fumigatus Δmsh6 Strains

2.4.1. Vector Construction

To assess the role of Msh6, the corresponding *msh6* gene was deleted from the *A. fumigatus* strain *aku*B^{KU80} and replaced by the resistance marker pyrithiamine (*prt*) (Figure 1). The corresponding fusion cassette was constructed by overlapping PCR [38]. The PCR conditions for amplification of upstream, downstream, and the *prt* fragment excerpts were 2 min at 98 °C and then 30 cycles of 30 s at 98 °C, 15 s at 55 °C and 2 min at 98 °C, with a final elongation step of 7 min at 68 °C. The PCR reaction mix was performed using the PrimeSTAR® HS DNA Polymerase (Takara Bio USA, Inc., San Jose, CA, USA) following the manufacturer's instructions using primers p1 and p2 for the upstream excerpt, p4 and p6 for the downstream and p3 and p5 fo the *prt* insert (Table S2). For the amplification of the final fusion cassette, PCR conditions were 95 °C for 1 min, then 35 cycles of 95 °C for 30 s, followed by 6 min at 68 °C, ending with a final step of 6 min at 68 °C. Primers used for amplifying the whole cassette were p1 and p6 (Table S2). PCR was performed using the Advantage® Polymerase Mix (Takara Bio USA, Inc.) following the manufacturer's recommendations, mixing equal amounts of the three excerpts to conform the cassette. The primer locations and the fusion vector design are shown in Figure 1.

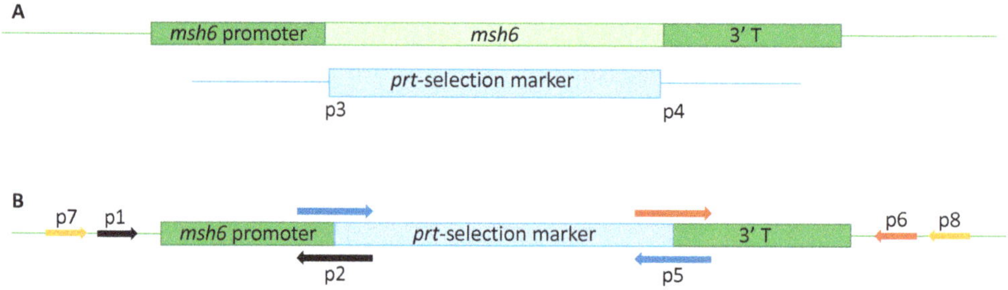

Figure 1. Construction of the fusion cassette for *A. fumigatus msh6* deletion in the *aku*B^{KU80} strain. The deleted *msh6* gene is indicated in light green, and the pyrithiamine selection marker in light blue. (**A**) Map of the parental *msh6*-WT strain and the pyrithiamine (*prt*) selection marker. (**B**) Design of the fusion cassette for *A. fumigatus* Δ*msh6* strain.

2.4.2. Transformation

Aspergillus fumigatus transformation was achieved using protoplasts, as described before [39]. The resulting transformants were selected on minimal medium plates containing 0.5 g/mL pyrithiamine (Sigma Aldrich, Madrid, Spain). Mutants were confirmed by PCR and sequencing using primers p7 and p8 (Figure 1, Table S3), and the positive ones were named with a letter T (transformant) followed by a number.

2.5. Phenotypic Characterization of Δmsh6 Mutant

A total of 10^4 conidia from two independent Δ*msh6* strains (T3 and T6) and its parental WT strain (*aku*B^{KU80}) were grown in solid MM containing 0.04 µM, 0.06 µM, and 1.6 µM of menadione; 0.005% and 0.02% of H_2O_2; 10 mg/L, 50 mg/L, and 100 mg/L of Congo Red (Sigma, Madrid, Spain); and 10 mg/L and 100 mg/L of Calcofluor white (Sigma, Madrid, Spain). Plates were incubated for 72 h at 37 °C, and at least two biological replicates were performed for each test.

2.6. Galleria Mellonella Survival Assay

To determine whether there were differences in virulence, the parental strain and the generated Δ*msh6* (T3 and T6) mutant strains were tested in the alternative model of infection *G. mellonella*. Wax-moth larvae used (TruLarv™ BioSystems Technology, Exeter, UK) were infected with *A. fumigatus* mutants and Δ*aku*B^{KU80} strain as a control. Wax-moth larvae killing assays were performed as previously described [40]. Twenty larvae (0.2–0.3 g) were inoculated with 10^3 and 10^4 conidia per larvae for each strain and incubated at 37 °C for 8 days, during which mortality was recorded daily. Each experiment was performed at least 3 times, and results were reported as mean values. Statistical analyses were performed with the GraphPad Prism software package (version 11.0) (SPSS Inc., Chicago, IL, USA). Kaplan-Meier survival curves were analyzed by using a log-rank (Mantel-Cox) test for significance. A p-value < 0.01 was considered significant.

2.7. Antifungal Susceptibility Testing

Antifungal susceptibility tests were performed following the E-test method [41] for itraconazole (ITC), voriconazole (VCZ), and Posaconazole (PSZ) (bioMérieux, Madrid, Spain). An inoculum adjusted to 10^5 conidia per ml was used, and 200 µL were plated in RPMI 1640 (Sigma-Aldrich Quimica SA, Madrid, Spain) agar base plates supplemented with 2% glucose. Minimal inhibitory concentrations (MICs) were read after 48 h incubation at 35 °C. Confirmation of antifungal susceptibility testing was performed using the broth microdilution method described by the European Committee on Antimicrobial Susceptibility Testing (EUCAST) [42]. The antifungal agents tested were ITC (Janssen Pharmaceutical, Madrid, Spain), VCZ (Pfizer, Madrid, Spain), PSZ (Schering-Plough, Madrid, Spain), and Amphotericin B (AmB) (Sigma, Madrid, Spain). *Aspergillus flavus* ATCC204304 and *A. fumigatus* ATCC204305 were used as quality control strains. In vitro susceptibility and resistance were defined according to the epidemiological cutoff values published for *A. fumigatus*. Isolates with ITC and VCZ MICs $\leq$ 1 µg/mL and PSZ $\leq$ 0.25 µg/mL were considered susceptible [43]. Both susceptibility tests were repeated at least twice for each strain.

2.8. Mutagenesis Experiments in Δmsh6 Strains

Putative gains of antifungal resistance were assayed by plating a total of 10^9 conidia of two Δ*msh6* null mutant strains (T3 and T6) and their parental WT strain (*aku*B^{KU80}) on solid minimal media (MM) supplemented with one of the following: 2 mg/L of PSZ, 4 mg/L of VCZ, 32 mg/L of prochloraz (PRZ), benomyl (BNY), boscalid (BCL), or azoxystrobin (AZB). All plates were incubated at 37 °C for up to 14 days until resistant colonies emerged. Colonies were picked and streaked in a new MM plate supplemented with the same antifungal at the matching concentration. Strains that grew again in these plates were picked and isolated for further genetic analysis.

2.9. Genetic Analysis of the Mutants

Genomic DNA was extracted from all *A. fumigatus* mutants, followed by PCR amplification and sequencing of *sdh*B, *ben*A, and *cyp*51A genes, as previously described [8,44].

3. Results

3.1. Analysis of Mutations in MMR Genes in a Collection of Isolates

We previously performed a whole genome sequencing analysis using a collection of 163 *A. fumigatus* genomes, from which a phylogenetic tree was built [8]. In that analysis, the strains were grouped in four different clusters in which azole-resistant strains were located in clusters I and II (Figure 2).

Azole-resistant strains with point mutations in *cyp*51A were distributed among clusters I and II, while TR resistance mechanisms were only located in three subclusters within cluster II that only contained TR strains. In the present study, the MMR genes *msh6*, *msh2*, *pms1*, and *mlh1* were analyzed in the *A. fumigatus* genome collection. The mutations detected in these genes are shown in Table 1, together with the percentage of strains harboring them. The analysis showed that genes *msh2*, *pms1*, and *mlh1* have low genetic variability and do not seem to correlate with azole drug resistance. Several point mutations were found in all genes. However, most of them were cluster-related and more likely representative of the clade.

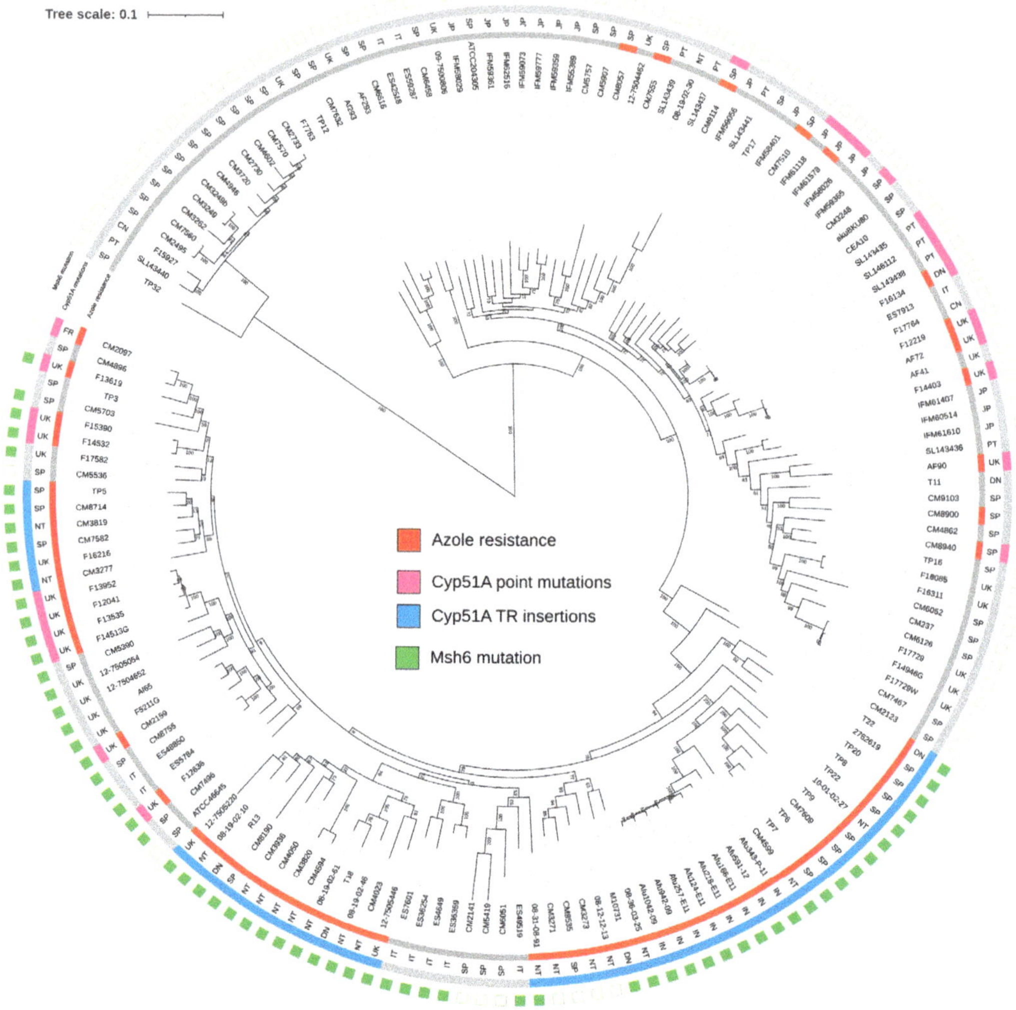

Figure 2. Phylogenetic tree derived from whole genome sequencing of 163 *A. fumigatus* strains. In blue, strains with TR_{34}/L98H mutations in Cyp51A; in green, Msh6 mutation (G240A).

Remarkably, a mutation in *msh6* leading to the substitution of glycine (G) for alanine (A) in position 240 (G240A) of the Msh6 protein was found in 42.8% of the isolates and only present in strains from cluster II. Moreover, almost all the strains harboring the TR_{34}/L98H azole resistance mechanism also had the G240A Msh6 mutation (Figure 2).

3.2. Construction and Phenotypic Characterization of *A. fumigatus* Δ*msh6* Strains

As such a high proportion of the G240A–Msh6 mutation was detected (42%), mainly in azole-resistant strains, we decided to investigate the potential contribution of *msh6* to the development of drug resistance in *A. fumigatus.* To this end, we deleted the entire *msh6* gene in the *aku*B^{KU80} strain. To eliminate the possibility of secondary mutations occurring during the construction of the knockout strains, we selected two independent biological replicate transformants to pursue all our phenotypic analyses (strains T3 and T6). To confirm the *msh6* gene deletion, the region was amplified and sequenced, confirming that the *ptrA* gene had replaced the *msh6* gene correctly. Both Δ*msh6* mutants were morphologically indistinguishable from the parental strain regarding macroscopic and microscopic morphology as well as colony radial growth), demonstrating that *msh6* is not required for normal growth.

Table 1. Mutations detected in the WGS analysis of *msh6*, *msh2*, *pms1*, and *mlh1* and the percentage of strains harboring them. Mutation G240A, in bold, was the most prevalent.

Gene (Gene Code)	Mutations	% of Strains
msh6 (Afu4g08300)	A55V	0.62
	V118A	0.62
	D121E	0.62
	G178A	1.86
	I183R	10.56
	G240A	**42.86**
	N289S	2.48
msh2 (Afu3g09850)	A45T	3.73
	P329T	3.73
	E467D	0.62
	E812G	1.24
	A889E	0.62
pms1 (Afu2g13410)	G286C	0.62
	P401A, V438A, K464R, Q611E, E87K, E760K	4.35
	E444G	2.48
	S758Y	1.24
	D1013Y	0.62
mlh1 (Afu5g11700)	K310R	4.35
	S368N	4.35
	I510T	1.86
	A641S	4.35

3.3. Deletion of msh6 Does Not Influence A. fumigatus Azole Susceptibility

E-tests were performed using ITC, VCZ, and PSZ in both the Δ*msh6* and the *aku*B^{KU80} strains (Figure S1). Microdilution susceptibility assays were also performed with ITC, VCZ, PSZ, and isavuconazole (IVZ). There was no difference in the MIC profile of the Δ*msh6* strains when compared with the parental strain (Table 2), showing that the deletion of *msh6*, in itself, has no direct impact on azole susceptibility.

Table 2. MIC ranges (EUCAST) to azole drugs for *aku*B^{KU80} and Δ*msh6 A. fumigatus* strains.

Strains	MIC Ranges (mg/L)			
	ITC	VCZ	PSZ	IVZ
*aku*B^{KU80}	1–1	0.5–0.5	0.125–0.25	1–1
Δ*msh6* T3	0.5–1	0.5–1	0.125–0.25	0.5–1
Δ*msh6* T6	0.5–1	0.5–0.5	0.25–0.25	1–1

3.4. Deletion of msh6 Does Not Influence A. fumigatus Virulence

To determine whether there were any differences in virulence between the WT strain and Δ*msh6* mutants, the *Galleria mellonella* alternative model of infection was used.

Using two different infection inocula, we observed that the mortality caused by both Δmsh6 mutants (T3 and T6) was the same as the mortality caused by the parental strain *aku*B^{KU80} (Figure 3). The inoculum of 10^4 conidia/larvae caused 100% mortality by day 5 post-infection for all strains (Figure 3 Left panel), and the inoculum of 10^3 conidia/larvae caused 80–90% mortality by day 8 post-infection for all strains (Figure 3 Right panel). The *A. fumigatus* Δ*msh6* mutants had similar virulence compared to their *akuB*^{KU80} parental strain with both inocula.

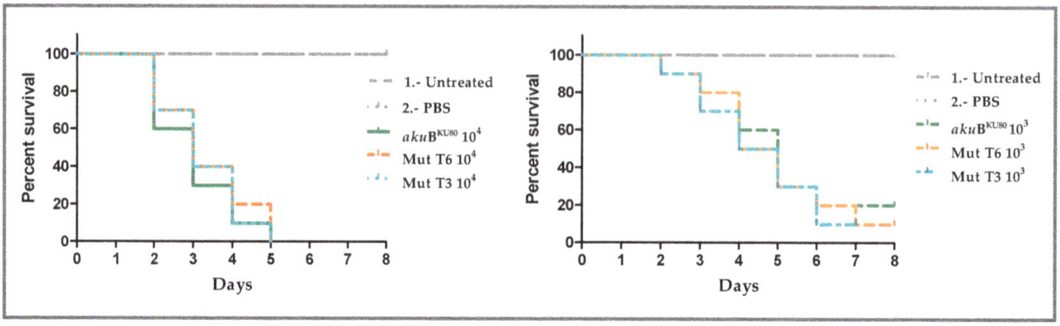

Figure 3. Survival curves of *G. mellonella* larvae inoculated with *aku*B^{KU80} and the Δ*msh6* (T3 and T6) *A. fumigatus* strains. Using 10^4 (**Left Panel**) and 10^3 (**Right Panel**) conidia/larvae inocula. Twenty larvae per isolate and group were used. Control groups include untreated larvae and larvae inoculated only with PBS. The experiment was repeated independently three times.

3.5. Deletion of msh6 Does Not Influence A. fumigatus Growth in Different Stress Conditions

Strains were subjected to growth at a range of different temperatures (37 °C, 45 °C, 50 °C and 60 °C), showing no differences between the parental and the Δ*msh6* strains. Strains were also subjected to oxidative stress with menadione at 0.04 mM and 0.05 mM, as well as cell wall stress with Congo Red at 10–25–50 µg/mL and Calcofluor white at 10–100 µg/mL. No differences were observed in any of the conditions tested.

3.6. Effects of Δmsh6 Deletion on A. fumigatus Antifungal Resistance Development

3.6.1. Development of Resistance to Azole Drugs

A total of 109 conidia of the parental strain (*aku*B^{KU80}) and the Δ*msh6* strains (T3 and T6) were plated on high concentrations of three azole drugs: VCZ, PSZ, and PRZ. Derived mutant colonies were only found on MM + 2 mg/L PSZ. There were no significant differences in the number of mutants obtained from each Δ*msh6* independent mutant (T3: 4.8 +/− 2.08; T6: 2.66 +/− 0.58) and the parental strain (3.75 +/− 4.5) in three independent experiments ($p = 0.8$, one way ANOVA con Dunnett's multiple comparisons). No mutants grew on any plate containing VCZ or PRZ.

3.6.2. Development of Resistance to Non-Azole Fungicides

Conidia of the parental strain (*aku*B^{KU80}) and both Δ*msh6* strains (T3 and T6) were plated on high concentrations of three non-azole fungicides commonly used in crop protection; they all belong to different antifungal families and target different genes.

BCL was selected as the representative drug for SDHIs, BNY for MBCs, and AZB for QoIs. Deletion of *msh6* in *A. fumigatus* led to the growth of significantly more resistant colonies after selection on BCL (WT: 5.8 +/− 4.85; T3: 70.1 +/− 20.48; T6: 56.6 +/− 7.23) and BNY (WT: 5.3 +/− 2.31; T3: 12 +/− 5.65; T6: 25 +/− 5.29). Antifungal selection revealed that the Δ*msh6* strains generated 10- and 2-fold more BCL- and BNY-resistant mutants, respectively, when compared with the parental strain (Figure 4). No resistant mutants were obtained on AZB.

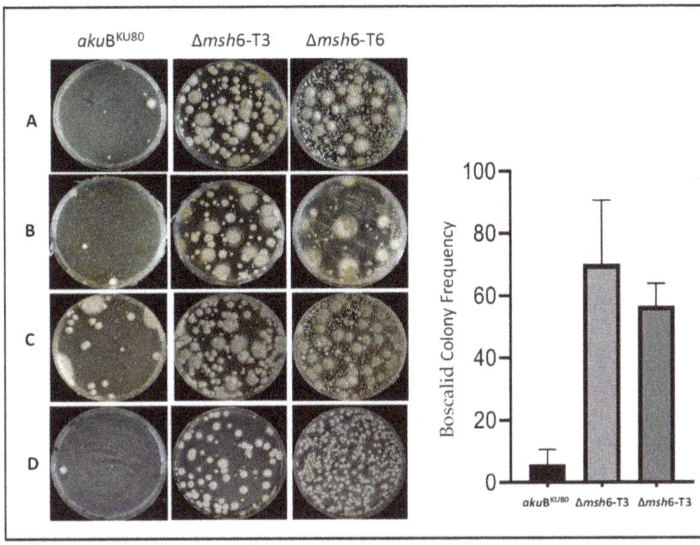

Figure 4. (**Left panel**) *aku*B^{KU80} and Δ*msh*6 mutant strains (1 × 10^9 conidia) after 32 mg/L BCL exposure in MM incubated for 7 days at 37 °C. Four independent experiments (**A–D**). (**Right panel**) CFUs grown in boscalid-supplemented media. $p < 0.001$ (one-way ANOVA with Dunnett's multiple comparisons).

3.7. Genetic Analysis of Drug Target Genes

Analysis of the *cyp*51A, *ben*A, and *shd*B polymorphisms and amino acid substitutions found in our chosen set of 123 *A. fumigatus* mutant strains are shown in Table 3. All PSZ mutants were subjected to PCR amplification and sequencing of *cyp*51A. Cyp51A analysis revealed the substitution of glycine (G) in position 54 for tryptophan (W) in 55 strains and arginine (R) in 1 strain. Both mutations have been previously associated with high resistance to long-tailed azole drugs, such as PSZ [45,46].

Table 3. Analysis of the *cyp*51A, *ben*A, and *shd*B polymorphisms and amino acid substitutions found in our set of 123 *A. fumigatus msh*6 mutant strains. Only polymorphisms involving non-synonymous mutations are shown. In bold, amino acid substitutions that have been previously found in resistant strains.

Gene	Nucleotide Position (cDNA)	Codon	Amino Acid Change	*aku*B^{KU80}	Δ*msh*6-T3	Δ*msh*6-T6
***cyp*51A**	g160t	Ggg/Tgg	**G54W**	16	31	80
	g160a	Ggg/Agg	**G54R**	0	1	0
***ben*A**	a593c	gAg/gCg	**E198A**	0	0	1
	g592c	Gag/Cag	**E198Q**	0	0	0
	g592a	Gag/Aag	E198K	0	8	0
	g594c	gag/gaC	E198D	0	3	3
	a593c	gAg/gTg	E198V	0	1	0
	a593g	gAg/gGg	E198G	2	0	0
	t599a	tTc/tCc	F200S	0	10	2
	t599a	tTc/tAc	**F200Y**	0	0	0
***sdh*B**	a809t	cAc/cTc	H270L	0	9	3
	c808t	Cac/Tac	**H270Y**	3	7	7
	a809g	cAc/cGc	**H270R**	0	8	0

Over 99% of boscalid selected colonies displayed a high level of resistance to BCL (MICs > 32 μg/mL), and 100% of them (37/37) had a mutation in the target gene (Table 3). PCR amplification and sequencing of these 37 BCL-resistant isolates revealed an amino acid substitution in SdhB of histidine (H) in position 270 for leucine (L) in 12 strains, tyrosine (Y) in 17 strains, and arginine (R) in 8 strains, which seems to be associated with resistance to BCL.

All benomyl-selected colonies displayed a high level of resistance to BNY (MICs > 32 μg/mL). PCR amplification and sequencing of the *ben*A gene were performed for all 30 *A. fumigatus* BNY-resistant strains, showing substitutions of a glutamic acid (E) for six different amino acids, including alanine (A) in one strain, lysine (K) harbored by eight strains, aspartic acid (D) in six strains, valine (V) in one strain or glycine (G) harbored by two strains. Lastly, the amino acid substitution of a phenylalanine (F) for a serine (S) in position 200 was present in 12 isolates.

4. Discussion

The worldwide emergence of *A. fumigatus* azole resistance is an important concern in the management of pathologies caused by this important fungal pathogen [3,9,47]. In the agricultural setting, *A. fumigatus*, together with other fungal plant pathogens, are exposed to the selective pressure of DMI drugs and to non-azole fungicides such as QoIs, SDHIs, or MBCs, used to protect crops. In this setting, multiresistant fungal strains are presumably being selected due to mutations on their respective targets, *cyp*51A, *cyt*B, sdhB, and *ben*A genes [8,14–17]. Based on a previous *A. fumigatus* WGS analysis [8], we found that all genomes harboring the TR_{34}/L98H Cyp51A alleles were grouped into three sub-clusters that included resistance to non-azole fungicides. We hypothesized that genome instability could be a potential cause for the development of azole resistance in *A. fumigatus*, related to the phenomenon of multi-drug resistance, as has been suggested before [19,20]. One of the pathways that recognize and repair mistakes and instabilities during DNA replication is the MMR system [26,48], a well-conserved pathway among the phylum *Ascomycota* [29]. It has been reported that deficiencies in MMR genes alter the proper function of this pathway, leading to errors in the reparation of mutations during cell replication [21]. In this study, we investigated four *A. fumigatus* MMR genes, *msh*2, *msh*6, *pms*1, and *mlh*1, to evaluate their possible implication in genetic instability and, thus, their potential contribution to the development of antifungal resistance. Among the MMR genes selected, *msh*6 stands out as the gene that harbors the most prevalent mutation, G240A, present in 42.8% of the strains analyzed. We observed *msh*6 variants in clinical and environmental *A. fumigatus* isolates investigated, which contrasts with the 18.2% *msh*2 variants found in *A. fumigatus* clinical isolates reported previously [20]. Interestingly, we observed that all *msh*6 variants grouped together within Cluster II and, moreover, most of these strains also harbored a TR_{34}/L98H azole resistance mechanism and several mutations leading to non-azole fungicide resistance [8]. Based on this, we speculated that mutations in *msh*6 could contribute, to some extent, to the development of antifungal drugs' multiresistance.

To evaluate the contribution of Msh6 with mutation development we deleted the *msh*6 gene in an *A. fumigatus* azole susceptible wild-type strain. Two Δ*msh*6 independent transformants, T3 and T6, were phenotypically characterized and compared to their parental strain (*aku*B^{KU80}). The phenotypic characterization performed did not disclose any difference in growth under various cell stresses, antifungal azole susceptibility, or virulence in the alternative *Galleria mellonella* animal model of infection. These results contrast with those previously obtained for *A. fumigatus msh*2 null mutants, which showed a reduction in virulence compared to the same parental strain [20]. However, our results are in consonance with studies in *Candida glabrata* and *Cryptococcus neoformans*, which demonstrated levels of colonization and infection of *msh*2 null mutants similar to those of the corresponding wild-type strains [21,49,50].

The deletion of *msh6* does not seem to cause a decrease in fitness. On the other hand, the abundance of strains found with mutations in *msh6* among clinical strains suggests that these changes could provide some benefits. Therefore, the potentially increased adaptative capacity to different situations, such as antifungal exposure or host immune system attack, could provide a benefit in certain conditions [28,51–54]. Because all *msh6* mutated strains were located in a specific cluster, we decided to investigate if an alteration of the MMR pathway functions could derive in a mutator phenotype with an increased mutation frequency in comparison with their parental strain [30].

Since the development of antifungal resistance involves a major problem in the management and treatment of fungal infections, the implication of hypermutator strains has been studied in different human pathogens, as well as the induction of antifungal resistance in in vitro conditions [20,55]. To evaluate the implications of the *msh6* deletion in antifungal resistance development, we grew Δ*msh6* isolates and their parental strain (*aku*B^{KU80}) under the pressure of azole and non-azole fungicides. Azole drug selection was derived in mutant colonies that grew only on agar plates supplemented with PSZ and not with VCZ or PCZ. However, the number of mutated isolates originating from *msh6* null mutants did not differ significantly from those of the wild-type strain. Previous studies in other fungal pathogens showed mixed results regarding the implication of MMR genes in the development of azole resistance. In *Candida albicans*, the lack of MSH2 protein has been associated with an increment in the frequency of fluconazole-resistant colony isolation [23]. However, the role of mutations in MSH2 in *C. glabrata* remains unclear, considering that several studies did not find an association between *C. glabrata* isolates harboring MSH2 mutations and the acquisition of azole or echinocandin resistance [50,56]. Nevertheless, other studies have reported the possible promotion of antifungal resistance due to the presence of mutations in MSH2 in *C. glabrata*. This could be possible if the MMR performed a limited and specific role in the acquisition of resistance to antifungals, as is the case of *C. glabrata* [49]. Concerning the genus *Cryptococcus*, a study on MMR in *C. neoformans* demonstrated that the deficiency of MSH2, MLH1, and PMS1 triggered a faster development of resistance to antifungals, as fluconazole-resistant colonies were isolated. However, the mechanisms of genetic stability in relation to azole resistance development have only started to be studied in *A. fumigatus*. To date, in the MMR pathway only Msh2 has been investigated, suggesting that the genetic instability caused by Δ*msh*A mutations can confer an adaptive advantage by increasing posaconazole resistance and virulence acquisition [20]. These same authors, exploring additional DDR mechanisms, have demonstrated the effect of deleting two kinases (*atm*A and *atr*A), conferring an advantage towards the intensity of voriconazole resistance acquisition [19]. Regardless of their individual implications in *A. fumigatus*, it still remains unknown if DNA damage repair and azole resistance development exist in a dynamic and interconnected network [57].

In our study, the analysis of mutation frequencies of *A. fumigatus* Δ*msh6* strains showed that the loss of *msh6* results in an increase in mutation numbers in genes that confer resistance to the non-azole fungicides BCL and BNY. The Δ*msh6* mutants produced 10 times more BCL-resistant isolates and two times more BNY-resistant isolates than the parental strain. Apart from the increased mutated frequencies obtained under this non-azole fungicide, a greater variety of amino acid substitutions were selected. Under BNY pressure, the most common β-tubulin *A. fumigatus* mutations, E198A/Q/and F200Y, were identified in the selected mutants, as well as other mutations such as E198K/D/V/G and F200S that have never been reported in *A. fumigatus* environmental isolates [8,11,12]. Similarly, under BCL pressure, the common H270Y/R *sdh*B field mutations were identified in the laboratory mutants, together with H270S that have never been reported in *A. fumigatus* environmental isolates [8,11]. This finding is in consonance with previous studies that report that laboratory mutagenesis generates a wide variety of mutations, of which only a subset will be reported in the field [58].

However, we did not find mutants resistant to azole drugs other than posaconazole, which makes us wonder whether this finding could be related to the conditions of the mutagenesis assay (very different from what happens in nature) or to additional mechanisms necessary to develop other mutations related to the azole drugs resistance. In addition, the multiple mutations in promoter and coding sequences of *cyp*51A responsible for multi-azole resistance (TR_{34}/L98H or TR_{46}/Y121F/T289A) likely occur in natural environments in a two or three-step sequential process, as it is known that single mutations (TR_{34} or L98H separately and TR_{46} or Y121F separately) slightly increase the MIC but do not confer complete resistance [44,59]. Therefore, it would not be possible to isolate these mutants in one step with the high level of drug used.

5. Conclusions

The results obtained in this study support the hypothesis that the acquisition of antifungal resistance in *A. fumigatus* could at least partially be driven by the genetic instability caused by alterations in the MMR. This hypothesis is also supported by similar findings reported in other fungal species, including *A. fumigatus* [20,21,23,49]. Hence, all evidence suggests that an *A. fumigatus* hypermutator phenotype could provide an advantage in drug-exposed environments without being detrimental to its fungal fitness or virulence.

Supplementary Materials: The following supporting information can be downloaded at: https://www.mdpi.com/article/10.3390/jof10030210/s1. Figure S1. Azole susceptibility tests (E-tests) were performed on the *aku*B^{KU80} strain and the Δ*msh6 A. fumigatus* strains T3 and T6. Table S1. Origin, year of isolation, and geographical origin of the strains used in the NGS analysis. Mutations of the genes described in this work; Table S2. Primers for the construction of the Δ*msh6* deletion cassette; Table S3. Primers used for sequencing of the Δ*msh6* deletion cassette.

Author Contributions: Conceptualization, E.M. and L.A.-F.; methodology, J.L., A.R. and I.G.-J.; software, validation, J.A. and L.A.-F.; formal analysis, I.G.-J., J.L. and L.A.-F.; investigation, I.G.-J. and J.L.; resources, E.M.; data curation, I.G.-J., J.A. and L.A.-F.; writing—original draft preparation, J.L., I.G.-J., L.A.-F. and E.M.; writing—review and editing, J.A., L.A.-F. and E.M.; visualization, all; supervision, E.M.; project administration, E.M.; funding acquisition, E.M. All authors have read and agreed to the published version of the manuscript.

Funding: This research was funded by Fondo de Investigacion Sanitaria (FIS PI21CIII/00028). J. Lucio held a predoctoral fellowship from the Fondo de Investigación Sanitaria (F17CIII/00037). This research is also supported by the Center for Biomedical Research in Infectious Diseases (CIBERINFEC-CB21/13/00105) Instituto de Salud Carlos III (ISCIII), Majadahonda, Madrid, Spain.

Data Availability Statement: Data are contained within the article and Supplementary Materials.

Conflicts of Interest: The authors declare that the research was conducted in the absence of any commercial or financial relationships that could be construed as a potential conflict of interest.

References

1. Patterson, T.F.; Thompson, G.R.; Denning, D.W.; Fishman, J.A.; Hadley, S.; Herbrecht, R.; Kontoyiannis, D.P.; Marr, K.A.; Morrison, V.A.; Nguyen, M.H.; et al. Practice guidelines for the diagnosis and management of aspergillosis: 2016 update by the infectious diseases society of America. *Clin. Infect. Dis.* **2016**, *63*, e1–e60. [CrossRef]
2. Verweij, P.E.; Ananda-Rajah, M.; Andes, D.; Arendrup, M.C.; Brüggemann, R.J.; Chowdhary, A.; Cornely, O.A.; Denning, D.W.; Groll, A.H.; Izumikawa, K.; et al. International expert opinion on the management of infection caused by azole-resistant *Aspergillus fumigatus*. *Drug Resist. Updates* **2015**, *21–22*, 30–40. [CrossRef]
3. Latgé, J.P.; Chamilos, G. *Aspergillus fumigatus* and aspergillosis in 2019. *Clin. Microbiol. Rev.* **2019**, *33*, e00140-18. [CrossRef]
4. Lepesheva, G.I.; Waterman, M.R. Sterol 14alpha-demethylase cytochrome P450 (CYP51), a P450 in all biological kingdoms. *Biochim. Biophys. Acta* **2007**, *1770*, 467–477. [CrossRef] [PubMed]
5. Garcia-Rubio, R.; Gonzalez-Jimenez, I.; Lucio, J.; Mellado, E. *Aspergillus fumigatus* Cross-Resistance between Clinical and Demethylase Inhibitor Azole Drugs. *Appl. Environ. Microbiol.* **2021**, *87*, e02539-20. [CrossRef] [PubMed]
6. Snelders, E.; Huis In 't Veld, R.A.G.; Rijs, A.J.M.M.; Kema, G.H.J.; Melchers, W.J.G.; Verweij, P.E. Possible environmental origin of resistance of *Aspergillus fumigatus* to medical triazoles. *Appl. Environ. Microbiol.* **2009**, *75*, 4053–4057. [CrossRef] [PubMed]
7. Zhang, J.; van den Heuvel, J.; Debets, A.J.M.; Verweij, P.E.; Melchers, W.J.G.; Zwaan, B.J.; Schoustra, S.E. Evolution of cross-resistance to medical triazoles in *Aspergillus fumigatus* through selection pressure of environmental fungicides. *Proc. R. Soc. B Biol. Sci.* **2017**, *284*, 20170635. [CrossRef]

8. Gonzalez-Jimenez, I.; Garcia-Rubio, R.; Monzon, S.; Lucio, J.; Cuesta, I.; Mellado, E. Multiresistance to nonazole fungicides in *Aspergillus fumigatus* TR_{34}/L98H Azole-resistant isolates. *Antimicrob. Agents Chemother.* **2021**, *65*, AAC0064221. [CrossRef] [PubMed]
9. Garcia-Rubio, R.; Cuenca-Estrella, M.; Mellado, E. Triazole Resistance in *Aspergillus* Species: An Emerging Problem. *Drugs* **2017**, *77*, 599–613. [CrossRef]
10. Chowdhary, A.; Sharma, C.; Meis, J.F. Azole-Resistant Aspergillosis: Epidemiology, Molecular Mechanisms, and Treatment. *J. Infect. Dis.* **2017**, *216* (Suppl. S3), S436–S444. [CrossRef]
11. Fraaije, B.; Atkins, S.; Hanley, S.; Macdonald, A.; Lucas, J. The Multi-Fungicide Resistance Status of *Aspergillus fumigatus* Populations in Arable Soils and the Wider European Environment. *Front. Microbiol.* **2020**, *11*, 599233. [CrossRef] [PubMed]
12. Kang, S.E.; Sumabat, L.G.; Melie, T.; Mangum, B.; Momany, M.; Brewer, M.T. Evidence for the agricultural origin of resistance to multiple antimicrobials in *Aspergillus fumigatus*, a fungal pathogen of humans. *G3 Genes Genomes Genet.* **2022**, *12*, jkab427. [CrossRef]
13. Hahn, M. The rising threat of fungicide resistance in plant *Pathogenic fungi*: *Botrytis* as a case study. *J. Chem. Biol.* **2014**, *7*, 133–141. [CrossRef] [PubMed]
14. Grasso, V.; Palermo, S.; Sierotzki, H.; Garibaldi, A.; Gisi, U. Cytochrome b gene structure and consequences for resistance to Qo inhibitor fungicides in plant pathogens. *Pest Manag. Sci.* **2006**, *62*, 465–472. [CrossRef] [PubMed]
15. Veloukas, T.; Kalogeropoulou, P.; Markoglou, A.N.; Karaoglanidis, G.S. Fitness and competitive ability of *Botrytis cinerea* field isolates with dual resistance to SDHI and QoI fungicides, associated with several sdhB and the cytb G143A mutations. *Phytopathology* **2014**, *104*, 347–356. [CrossRef] [PubMed]
16. Sierotzki, H.; Scalliet, G. A review of current knowledge of resistance aspects for the next-generation succinate dehydrogenase inhibitor fungicides. *Phytopathology* **2013**, *103*, 880–887. [CrossRef]
17. Vela-Corcía, D.; Romero, D.; de Vicente, A.; Pérez-García, A. Analysis of β-tubulin-carbendazim interaction reveals that binding site for MBC fungicides does not include residues involved in fungicide resistance. *Sci. Rep.* **2018**, *8*, 7161. [CrossRef]
18. Garcia-Rubio, R.; Monzon, S.; Alcazar-Fuoli, L.; Cuesta, I.; Mellado, E. Genome-wide comparative analysis of *Aspergillus fumigatus* strains: The reference genome as a matter of concern. *Genes* **2018**, *9*, 363. [CrossRef]
19. dos Reis, T.F.; Silva, L.P.; de Castro, P.A.; Almeida de Lima, P.B.; do Carmo, R.A.; Marini, M.M.; da Silveira, J.F.; Ferreira, B.H.; Rodrigues, F.; Malavazi, I.; et al. The Influence of Genetic Stability on *Aspergillus fumigatus* Virulence and Azole Resistance. *G3 Genes Genomes Genet.* **2018**, *8*, 265–278. [CrossRef]
20. dos Reis, T.F.; Silva, L.P.; de Castro, P.A.; do Carmo, R.A.; Marini, M.M.; da Silveira, J.F.; Ferreira, B.H.; Rodrigues, F.; Lind, A.L.; Rokas, A.; et al. The *Aspergillus fumigatus* Mismatch Repair MSH2 Homolog Is Important for Virulence and Azole Resistance. *mSphere* **2019**, *4*, e00416-19. [CrossRef]
21. Boyce, K.J.; Wang, Y.; Verma, S.; Shakya VP, S.; Xue, C.; Idnurm, A. Mismatch repair of DNA replication errors contributes to microevolution in the pathogenic fungus *Cryptococcus neoformans*. *mBio* **2017**, *8*, 10–1128. [CrossRef]
22. Byun, S.A.; Won, E.J.; Kim, M.N.; Lee, W.G.; Lee, K.; Lee, H.S.; Uh, Y.; Healey, K.R.; Perlin, D.S.; Choi, M.J.; et al. Multilocus sequence typing (MLST) genotypes of *Candida glabrata* bloodstream isolates in Korea: Association with antifungal resistance, mutations in mismatch repair gene (Msh2), and clinical outcomes. *Front. Microbiol.* **2018**, *9*, 1523. [CrossRef]
23. Legrand, M.; Chan, C.L.; Jauert, P.A.; Kirkpatrick, D.T. Role of DNA mismatch repair and double-strand break repair in genome stability and antifungal drug resistance in *Candida albicans*. *Eukaryot. Cell* **2007**, *6*, 2194–2205. [CrossRef] [PubMed]
24. Chatterjee, N.; Walker, G.-C. Mechanisms of DNA damage, repair, and mutagenesis. *Environ. Mol. Mutagen.* **2017**, *58*, 235–263. [CrossRef] [PubMed]
25. Yao, S.; Feng, Y.; Zhang, Y.; Feng, J. DNA damage checkpoint and repair: From the budding yeast *Saccharomyces cerevisiae* to the pathogenic fungus *Candida albicans*. *Comput. Struct. Biotechnol. J.* **2021**, *19*, 6343–6354. [CrossRef]
26. Iyer, R.R.; Pluciennik, A.; Burdett, V.; Modrich, P.L. DNA Mismatch Repair: Functions and Mechanisms. *Chem. Rev.* **2006**, *106*, 302–323. [CrossRef] [PubMed]
27. Larrea, A.A.; Lujan, S.A.; Kunkel, T.A. SnapShot: DNA Mismatch Repair. *Cell* **2010**, *141*, 730. [CrossRef]
28. Boyce, K.J. Mutators Enhance Adaptive Micro-Evolution in Pathogenic Microbes. *Microorganisms* **2022**, *10*, 442. [CrossRef]
29. Phillips, M.A.; Steenwyk, J.L.; Shen, X.X.; Rokas, A. Examination of Gene Loss in the DNA Mismatch Repair Pathway and Its Mutational Consequences in a Fungal Phylum. *Genome Biol. Evol.* **2021**, *13*, evab219. [CrossRef]
30. Gambhir, N.; Harris, S.D.; Everhart, S.E. Evolutionary Significance of Fungal Hypermutators: Lessons Learned from Clinical Strains and Implications for Fungal Plant Pathogens. *mSphere* **2022**, *7*, e0008722. [CrossRef]
31. Alcazar-Fuoli, L.; Mellado, E.; Alastruey-Izquierdo, A.; Cuenca-Estrella, M.; Rodriguez-Tudela, J.L. *Aspergillus* section *Fumigati*: Antifungal susceptibility patterns and sequence-based identification. *Antimicrob. Agents Chemother.* **2008**, *52*, 1244–1251. [CrossRef]
32. Bolger, A.M.; Lohse, M.; Usadel, B. Trimmomatic: A flexible trimmer for Illumina sequence data. *Bioinformatics* **2014**, *30*, 2114–2120. [CrossRef]
33. Langmead, B.; Salzberg, S.L. Fast gapped-read alignment with Bowtie2. *Nat. Methods* **2012**, *9*, 357–359. [CrossRef] [PubMed]
34. Quinlan, A.R.; Hall, I.M. BEDTools: A flexible suite of utilities for comparing genomic features. *Bioinformatics* **2010**, *26*, 841–842. [CrossRef] [PubMed]

35. McKenna, A.; Hanna, M.; Banks, E.; Sivachenko, A.; Cibulskis, K.; Kernytsky, A.; Garimella, K.; Altshuler, D.; Gabriel, S.; Daly, M.; et al. The Genome Analysis toolkit: A MapReduce framework for analyzing next-generation DNA sequencing data. *Genome Res.* **2010**, *20*, 1297–1303. [CrossRef] [PubMed]
36. Stamatakis, A. RAxML version 8: A tool for phylogenetic analysis and post-analysis of large phylogenies. *Bioinformatics* **2014**, *30*, 1312–1313. [CrossRef] [PubMed]
37. Yu, G.; Smith, D.K.; Zhu, H.; Guan, Y.; Lam, T.T.Y. ggtree: An R package for visualization and annotation of phylogenetic trees with their covariates and other associated data. *Methods Ecol. Evol.* **2017**, *8*, 28–36. [CrossRef]
38. Alcazar-Fuoli, L.; Clavaud, C.; Lamarre, C.; Aimanianda, V.; Seidl-Seiboth, V.; Mellado, E.; Latgé, J.P. Functional analysis of the fungal/plant class chitinase family in *Aspergillus fumigatus*. *Fungal Genet. Biol.* **2011**, *48*, 418–429. [CrossRef]
39. Szewczyk, E.; Nayak, T.; Oakley, C.E.; Edgerton, H.; Xiong, Y.; Taheri-Talesh, N.; Osmani, S.A.; Oakley, B.R. Fusion PCR and gene targeting in *Aspergillus nidulans*. *Nat. Protoc.* **2006**, *1*, 3111–3120. [CrossRef]
40. Gomez-Lopez, A.; Forastiero, A.; Cendejas-Bueno, E.; Gregson, L.; Mellado, E.; Howard, S.J.; Livermore, J.L.; Hope, W.W.; Cuenca-Estrella, M. An invertebrate model to evaluate virulence in *Aspergillus fumigatus: The* role of azole resistance. *Med. Mycol.* **2014**, *52*, 311–319. [CrossRef]
41. Pfaller, J.B.; Messer, S.A.; Hollis, R.J.; Diekema, D.J.; Pfaller, M.A. In vitro susceptibility testing of *Aspergillus* spp.: Comparison of Etest and reference microdilution methods for determining voriconazole and itraconazole MICs. *J. Clin. Microbiol.* **2003**, *41*, 1126–1129. [CrossRef] [PubMed]
42. Guinea, J.; Meletiadis, J.; Arikan-Akdagli, S.; Kahlmeter, G.; Arendrup, M.C.; The Subcommittee on Antifungal Susceptibility Testing (AFST) of the ESCMID European Committee for Antimicrobial Susceptibility Testing (EUCAST). EUCAST DEFINITIVE DOCUMENT E.DEF 9.4 Method for the Determination of Broth Dilution Minimum Inhibitory Concentrations of Antifungal Agents for Conidia Forming Moulds. 2022. Available online: https://www.eucast.org/fileadmin/src/media/PDFs/EUCAST_files/AFST/Files/EUCAST_EDef_9.4_method_for_susceptibility_testing_of_moulds.pdf (accessed on 1 April 2022).
43. The European Committee on Antimicrobial Susceptibility Testing. Breakpoint Tables for Interpretation of MICs for Antifungal Agents, Version 10.0. 2020. Available online: http://www.eucast.org/astoffungi/clinicalbreakpointsforantifungals (accessed on 4 February 2020).
44. Mellado, E.; Garcia-Effron, G.; Alcázar-Fuoli, L.; Melchers WJ, G.; Verweij, P.E.; Cuenca-Estrella, M.; Rodríguez-Tudela, J.L. A new *Aspergillus fumigatus* resistance mechanism conferring in vitro cross-resistance to azole antifungals involves a combination of cyp51A alterations. *Antimicrob. Agents Chemother.* **2007**, *51*, 1897–1904. [CrossRef]
45. Mann, P.A.; Parmegiani, R.M.; Wei, S.Q.; Mendrick, C.A.; Li, X.; Loebenberg, D.; DiDomenico, B.; Hare, R.S.; Walker, S.S.; McNicholas, P.M. Mutations in *Aspergillus fumigatus* resulting in reduced susceptibility to posaconazole appear to be restricted to a single amino acid in the cytochrome P450 14alpha-demethylase. *Antimicrob. Agents Chemother.* **2003**, *47*, 577–581. [CrossRef] [PubMed]
46. Espinel-Ingroff, A.; Turnidge, J.; Alastruey-Izquierdo, A.; Dannaoui, E.; Garcia-Effron, G.; Guinea, J.; Kidd, S.; Pelaez, T.; Sanguinetti, M.; Meletiadis, J.; et al. Posaconazole MIC Distributions for *Aspergillus fumigatus* Species Complex by Four Methods: Impact of cyp51A Mutations on Estimation of Epidemiological Cutoff Values. *Antimicrob. Agents Chemother.* **2018**, *62*, e01916-17. [CrossRef]
47. Warris, A. Azole-resistant aspergillosis. *J. Infect.* **2015**, *71*, S121–S125. [CrossRef] [PubMed]
48. Gupta, D.; Heinen, C.D. The mismatch repair-dependent DNA damage response: Mechanisms and implications. *DNA Repair* **2019**, *78*, 60–69. [CrossRef] [PubMed]
49. Healey, K.R.; Zhao, Y.; Perez, W.B.; Lockhart, S.R.; Sobel, J.D.; Farmakiotis, D.; Kontoyiannis, D.P.; Sanglard, D.; Taj-Aldeen, S.J.; Alexander, B.D.; et al. Prevalent mutator genotype identified in fungal pathogen *Candida glabrata* promotes multi-drug resistance. *Nat. Commun.* **2016**, *7*, 11128. [CrossRef]
50. Bordallo-Cardona, M.Á.; Agnelli, C.; Gómez-Nuñez, A.; Sánchez-Carrillo, C.; Bouza, E.; Muñoz, P.; Escribano, P.; Guinea, J. MSH2 Gene Point Mutations Are Not Antifungal Resistance Markers *in Candida glabrata*. *Antimicrob. Agents Chemother.* **2019**, *63*, e01876-18. [CrossRef]
51. Billmyre, R.B.; Clancey, S.A.; Heitman, J. Natural mismatch repair mutations mediate phenotypic diversity and drug resistance in *Cryptococcus deuterogattii*. *eLife* **2017**, *6*, e28802. [CrossRef]
52. Buscaino, A. Chromatin-mediated regulation of genome plasticity in human fungal pathogens. *Genes* **2019**, *10*, 855. [CrossRef]
53. Sundin, G.W.; Weigand, M.R. The microbiology of mutability. *FEMS Microbiol. Lett.* **2007**, *277*, 11–20. [CrossRef] [PubMed]
54. Verweij, P.E.; Zhang, J.; Debets, A.J.M.; Meis, J.F.; van de Veerdonk, F.L.; Schoustra, S.E.; Zwaan, B.J.; Melchers, W.J.G. In-host adaptation and acquired triazole resistance in *Aspergillus fumigatus*: A dilemma for clinical management. *Lancet Infect. Dis.* **2016**, *16*, e251–e260. [CrossRef] [PubMed]
55. Hokken, M.W.J.; Zwaan, B.J.; Melchers, W.J.G.; Verweij, P.E. Facilitators of adaptation and antifungal resistance mechanisms in clinically relevant fungi. *Fungal Genet. Biol.* **2019**, *132*, 103254. [CrossRef] [PubMed]
56. Singh, A.; Healey, K.R.; Yadav, P.; Upadhyaya, G.; Sachdeva, N.; Sarma, S.; Kumar, A.; Tarai, B.; Perlin, D.S.; Chowdhary, A. Absence of azole or echinocandin resistance in *Candida glabrata* isolates in India despite background prevalence of strains with defects in the dna mismatch repair pathway. *Antimicrob. Agents Chemother.* **2018**, *62*, e00195-18. [CrossRef]
57. Peng, G.; Lin, S.Y. Exploiting the homologous recombination DNA repair network for targeted cancer therapy. *World J. Clin. Oncol.* **2011**, *2*, 73–79. [CrossRef]

58. Hawkins, N.J.; Fraaije, B.A. Predicting Resistance by Mutagenesis: Lessons from 45 Years of MBC Resistance. *Front. Microbiol.* **2016**, *7*, 1814. [CrossRef]
59. Snelders, E.; Camps, S.M.; Karawajczyk, A.; Rijs, A.J.; Zoll, J.; Verweij, P.E.; Melchers, W.J. Genotype-phenotype complexity of the TR46/Y121F/T289A cyp51A azole resistance mechanism in *Aspergillus fumigatus*. *Fungal Genet. Biol.* **2015**, *82*, 129–135. [CrossRef]

Journal of
Fungi

MDPI

Article

Comparative Genomics of the First Resistant *Candida auris* Strain Isolated in Mexico: Phylogenomic and Pan-Genomic Analysis and Mutations Associated with Antifungal Resistance

Arturo Casimiro-Ramos [1], Celia Bautista-Crescencio [1], Alvaro Vidal-Montiel [1], Gloria M. González [2], Juan Alfredo Hernández-García [1], César Hernández-Rodríguez [1] and Lourdes Villa-Tanaca [1,*]

1. Laboratorio de Biología Molecular de Bacterias y Levaduras, Departamento de Microbiología, Escuela Nacional de Ciencias Biológicas, Instituto Politécnico Nacional, Prolongación de Carpio y Plan de Ayala, Casco de Santo Tomás, Ciudad de México 11340, Mexico; acasimiror1800@alumno.ipn.mx (A.C.-R.); luna_20_08@hotmail.com (C.B.-C.); alvaro301013@gmail.com (A.V.-M.); freddyhardcore@gmail.com (J.A.H.-G.); chdez38@hotmail.com (C.H.-R.)
2. Departamento de Microbiología, Facultad de Medicina, Universidad Autónoma de Nuevo León, Hospital Universitario "Dr. José Eleuterio Gonzalez", Av. Madero y Calle Dr. Eduardo Aguirre Pequeño s/n, Colonia Mitras Centro, Monterrey 64460, Nuevo Leon, Mexico; gloria.gonzalezgn@uanl.edu.mx

* Correspondence: mvillat@ipn.mx; Tel.: +52-55-5729-63-00 (ext. 62569)

Abstract: *Candida auris* is an emerging multidrug-resistant and opportunistic pathogenic yeast. Whole-genome sequencing analysis has defined five major clades, each from a distinct geographic region. The current study aimed to examine the genome of the *C. auris* 20–1498 strain, which is the first isolate of this fungus identified in Mexico. Based on whole-genome sequencing, the draft genome was found to contain 70 contigs. It had a total genome size of 12.86 Mbp, an N50 value of 1.6 Mbp, and an average guanine-cytosine (GC) content of 45.5%. Genome annotation revealed a total of 5432 genes encoding 5515 proteins. According to the genomic analysis, the *C. auris* 20–1498 strain belongs to clade IV (containing strains endemic to South America). Of the two genes (*ERG11* and *FKS1*) associated with drug resistance in *C. auris*, a mutation was detected in K143R, a gene located in a mutation hotspot of *ERG11* (lanosterol 14-α-demethylase), an antifungal drug target. The focus on whole-genome sequencing and the identification of mutations linked to the drug resistance of fungi could lead to the discovery of new therapeutic targets and new antifungal compounds.

Keywords: *Candida auris*; whole genome sequencing (WGS); phylogenomics; pan-genome analysis; azole resistance; K143R; Erg 11 mutation

1. Introduction

Candida auris was first reported in Japan in 2009 after being isolated from the secretion of the external ear canal of a female patient [1]. Subsequently, clinical isolates of the same strain in South Korea were retrospectively identified, dating back to 1996 [2]. In a little over a decade, *C. auris* has emerged in healthcare settings worldwide and is suggested to be involved in numerous medical conditions. It is capable of colonizing the skin and causing outbreaks of invasive candidiasis. Bloodstream infections are the most frequent invasive condition and have been responsible for increasing in-hospital mortality rates by up to 70% [3,4].

As part of the public health response, whole-genome sequencing (WGS) has played a significant role in characterizing the transmission dynamics of *C. auris* and in detecting new outbreaks [3]. With this technique, it has been possible to define five clades and one potential sixth clade of *C. auris*. Clade I is manifested in South Asia, II in East Asia, III in Africa, IV in South America, and V in Iran [4–6]. A possible clade VI has been proposed based on three *C. auris* isolates found in Singapore that are genetically distinct from clades I–V [7].

Each clade has a specific level of resistance to the main antifungals administered to treat candidiasis (azoles, echinocandins, and amphotericin B). Isolates belonging to clade II are typically susceptible to azoles. At the same time, resistance to these drugs is shown by nearly all the isolates in clades I and III and by roughly half of those in clade IV. Some of the resistant isolates contain one of three mutations (F126L, Y132F, and K143R) in lanosterol 14-α-demethylase (ERG11), a drug target [3,4,8,9].

A small percentage of the isolates from clades I, III, and IV are resistant to echinocandins. Resistance has been linked to a single mutation at S639 (S639Y/P/F) in the hotspot 1 region of 1,3-β-D-glucan synthase (Fks1), another drug target [3,9]. Although low susceptibility to amphotericin B is common in clades I and IV, the mutations responsible for drug-resistance have not been defined [8]. In Mexico, *C. auris* 20–1498 was first isolated from a blood sample of a patient with gastrointestinal complications and endometriosis in May of 2020. This strain was identified by sequence analysis of the ITS1-5.8S-ITS2 and D1/D2 ribosomal regions [10].

The current contribution aimed to determine the genome of the *C. auris* 20–1498 strain, confirm the clade to which it belongs by comparative genomics, and analyze the mutations responsible for its resistance to azoles. Information on the genome of this strain will allow for its comparison with the genome of the *C. auris* strain isolated months later in the same hospital and associated with COVID-19.

2. Materials and Methods

2.1. Clinical Data and Isolates

The *C. auris* 20–1498 strain was isolated from a blood culture of a 58-year-old woman with severe endometriosis (stage IV). The patient had a history of multiple hospitalizations and the application of a central catheter and Mahurkar catheter. The *C. auris* 20–1498 isolate was identified by the Microbiology Department of the Faculty of Medicine in the Universidad Autónoma de Nuevo Leon in conjunction with the University Hospital (Hospital Universitario Dr. Jose Eleuterio Gonzalez) [10].

2.2. Fungal Growth Conditions

The *C. auris* 20–1498 isolate was cultured on Sabouraud dextrose agar plates (SDA; MCD LAB, S.A. de C.V., Mexico) at 35 °C for 2 days. The morphology and purity of the *C. auris* 20–1498 colony were established on solid SDA.

2.3. DNA Extraction and Genome Sequencing

For the genomic identification of the *C. auris* 20–1498 isolate, DNA was extracted with the Zymo Research® Soil Microbe DNA Miniprep kit, Irvine, CA, USA. Whole-genome sequencing was carried out on the Illumina HiSeq 4000 system (Novogene, Sacramento, CA, USA).

2.4. Thermotolerance and Halotolerance

The thermotolerance of *C. auris* 20–1498 growth was tested according to the modified protocol described by Reséndiz-Sánchez et al. in 2020, using *C. haemulonii* 87, *C. albicans* ATCC 10231, *C. glabrata* CBS 138, and *C. auris* CJ97 as controls. The yeasts were grown in yeast extract peptone dextrose (YPD) broth under constant shaking at 28 °C until reaching the early stationary growth phase (~15 h). The inoculum was adjusted to $As_{600} = 0.5$ with sterile YPD medium, and 5 μL of each strain was inoculated into the corresponding culture medium and streaked with a microbiological loop. The cultures were incubated at different temperatures (28, 37, and 42 °C), and yeast growth was recorded every 24 h for 3 days. The solid culture media utilized were SDA, YPD, YPD with 1 M NaCl, YPD with 2 M NaCl, and blood agar [11]. *C. haemulonii* was sensitive at temperatures of 37 and 42 °C in SDA medium. In contrast, *C. albicans* and *C. auris* could grow at all three tested temperatures.

2.5. Genome Assembly and Annotation

The whole-genome sequencing reads were assessed for quality with the FastQC v0.11.9 program [12] and then trimmed with Trimmomatic v0.39 [13]. Subsequently, the genome was assembled on Velvet v1.2.10 software by means of the referenced assembly method, with the *C. auris* B11220 strain (GCA_003013715.2) as the reference genome [14]. Finally, the assembly quality was evaluated with the QUAST v5.0.2 program [15]. The genome annotation of *C. auris* 20–1498 was achieved on the Companion v1.0.2 server [16], using *C. auris* B8441 as the reference strain to standardize the models for gene finding, functional annotation transfer, and pseudochromosome contiguation. The *C. auris* 20–1498 genome sequence and gene annotation generated were deposited in the GenBank BioProject (access number: PRJNA1013603) and BioSample ID (SAMN39051800).

2.6. Phylogenomic Tree and Pan-Genome of Candida auris

The phylogenomic tree contains a total of twenty-six *C. auris* yeast genomes. The genome sequences of the *C. auris* isolates were downloaded from the NCBI database [17]: B11205 (GCA_016772135.1), B8441 (GCA_002759435.2), B13916 (GCA_016772235.1), 20–26 (GCA_025429755.1), 20–32 (GCA_025429595.1), BJCA001 (GCA_018831645.1), CA8LBN (GCA_019039635.1), CA27LBN (GCA_019039335.1), L1537/2020 (GCA_020809265.1), RCPF-1821 (GCA_004287075.1), B12043 (GCA_016495645.1), B11809 (GCA_016495685.1), B13463 (GCA_016495665.1), B11220 (GCA_003013715.2), B11221 (GCA_002775015.1), B12037 (GCA_016772215.1), B12631 (GCA_016772195.1), BJCA002 (GCA_018902005.1), LOM (GCA_005234155.1), A1 (GCA_014217455.1), B17721 (GCA_016772175.1), B11243 (GCA_003014415.1), B11245 (GCA_008275145.1), B12342 (GCA_016772155.1), and IFRC2087 (GCA_016809505.1). Gene annotation was performed on the Companion server (http://companion.sanger.ac.uk, accessed on 9 March 2024). A maximum likelihood phylogenomic tree was constructed with OrthoFinder v4.0 software, utilizing a core-proteome-based phylogenomic analysis (CPBP) to obtain the clusters of orthologous groups of proteins (COGs) from the proteome of each organism [18].

The pan-genome analysis, encompassing the core genome, character genome, and accessory genome, was based on the annotated amino acid sequences of the following isolates: B8441, B11220, B11221, IFRC2087, B11243, B11245, B12342, and 20–1498. The pan-genome was constructed on the Orthovenn3 server [19], which incorporates the OrthoFinder algorithm. The phylogenomic tree was edited with Interactive Tree of Life v5 (iTOL) [20].

2.7. Phylogenetic and Comparative Analysis of the Erg11 and Fks1 Proteins

The phylogenetic trees were constructed for the Erg11 and Fks1 proteins based on their amino acid sequences. These were downloaded from the NCBI database and were comprised of different species of *Candida*, *Saccharomyces cerevisiae* S288C, and *Yarrowia lipolytica* CLIB122, which served as the outgroup species. The access numbers for Erg11 were the following: for *Candida auris*, 20–26 (000164700.1*), 20–32 (000136700.1*), B11220 (QEO20389.1), B11809 (000072800.1*), A1 (000324100.1*), LOM (000326000.1*), 20–1498 (000047500.1*), B11243 (PSK75255.1), B11245 (QEL61552.1), and IFRC2087 (QRG39199.1); for *Candida albicans*, SC5314 (XP_716761.1); for *Candida tropicalis*, MYA 3404 (XP_002550985.1); for *Candida parapsilosis*, ATCC 22019 (ACT67904.1); for *Candida lusitaniae*, ATCC 42720 (XP_002614916.1) and CBS 6936 (OVF10151.1); for *Candida haemulonii*, B11899 (XP_025344294.1) and LIP Ch2 (QOU12110.1); for *Candida duobushaemulonii*, B09383 (XP_025336625.1) and LIP Ch8 (QOU12108.1); for *Candida pseudohaemulonii*, B12108 (XP_024711630.1); for *Candida glabrata*, CBS 138 (XP_445876.1); for *Saccharomyces cerevisiae*, S288C (NP_011871.1); and for *Yarrowia lipolytica*, CLIB122 (XP_500518.1). The access numbers for Fks1 were the following: for *Candida auris*, 20–26 (000117700.1*), 20–32 (000098200.1*), B13916 (000299100.1*), B11220 (QEO20537.1), B11809 (000326000.1*), A1 (000218400.1*), LOM (000219600.1*), 20–1498 (000518900.1*), B11243 (PSK74959.1), and IFRC2087 (QRG37633.1); for *Candida albicans*, SC5314 (XP_721429.2); for *Candida tropicalis*, ATCC 750 (ACF22801.2); for *Candida orthopsilosis*, Co 90–125 (XP_003867907.1);

for *Candida lusitaniae*, L17 (KAF5212065.1) and P5 (QFZ43597.1); for *Candida haemulonii*, B11899 (XP_025339819.1); for *Candida duobushaemulonii*, B09383 (XP_025335019); for *Candida pseudohaemulonii*, B12108 (XP_024714055.1); for *Candida glabrata*, M6 (KAI8387093.1); for *Saccharomyces cerevisiae*, S288C (NP_013446.1); and for *Yarrowia lipolytica*, CLIB122 (XP_504213.2). The alignment was generated by the CLUSTAL W v7 program available via MEGA7 software, and the phylogram was constructed with the maximum likelihood method and the Le and Gascuel + G model available via MEGA 7 software [21] by performing 1000 bootstrap replicates. The phylogenetic trees of the Erg11 and Fks1 proteins were edited with iTOL [20]. The antifungal profile and point mutations of the isolates used in the analysis and the phylogenetic trees of the *C. auris* Erg11 and Fks1 proteins are shown in Table S1 (Supplementary Material).

2.8. Modeling the Erg11 Protein from Candida auris 20–1498

The 3D structure of the *Candida auris* 20–1498 Erg11 protein was initially generated with the homology modeling technique, employing the Erg11 protein of *Candida albicans* with RCSB PDB ID: 5fsa as the template [22]. The analysis was conducted using Modeller v10.1 software [23], which is widely used to generate comparative models. Ten models were generated, and the best model was selected based on the lowest molpdf score. The results were validated with a Ramachandran plot, finding a greater percentage of residues located in favorable regions [24].

2.9. Molecular-Docking Study of Some Azoles on the Erg11 Protein

To explore the affinity of lanosterol, mevalonate, and some azoles for Erg11, docking simulations were carried out on the 3D structure of the *Candida auris* 20–1498 Erg11 protein. The lanosterol substrate of the Erg11 enzyme served as the positive control and mevalonate as the negative control. Subsequently, the affinity of some azoles (e.g., fluconazole and voriconazole) for Erg11 was examined. The docking simulations were performed using Autodock vina v4.2 [25].

3. Results

3.1. C. auris 20–1498 Genome Assembly and Annotation

The sequencing of the genome of *C. auris* 20–1498 showed a draft genome of 70 contigs, a total genome size of 12.86 Mbp, an N50 value of 1.6 Mbp, and an average guanine-cytosine (GC) content of 45.5% (Table 1).

Table 1. Summary of the *C. auris* 20–1498 annotated genome assembly.

Features	Values
Quality of reads	QC [1] > 30
Genome assembly size (bp)	12,869,713
Number of contigs	70
N50 (bp)	1,604,526
Genome coverage	99.68
GC content (%)	45.5
Number of predicted genes	5432
Number of coding genes	5304
Number of genes with multiple CDSs	565
Number of hypothetical proteins	5515

[1] Quality control. GC, guanine-cytosine; CDS, coding sequences.

3.2. Phylogenetic Tree and Pan-Genome of Candida auris

To examine the phylogenetic relationship between *C. auris* 20–1498 and 25 strains of *C. auris* from different clades, single-copy orthologs in 26 sequenced genomes were identified by using OrthoFinder v2.5.4, which assigned 141,635 genes (99.8% of the total) to 5501 orthogroups. A total of 50% of all the genes were in orthogroups with 26 or more genes (G_{50} = 26) and were contained in the largest 2685 orthogroups (O_{50} = 2685). There were

5014 orthogroups among all the species, and 4723 of these consisted entirely of single-copy genes. Of the five known clades of *C. auris* in the world [4,5], the whole-genome sequencing of the isolate from Mexico confirmed that it was genetically closest to clade IV (containing strains endemic to South America) (Figure 1).

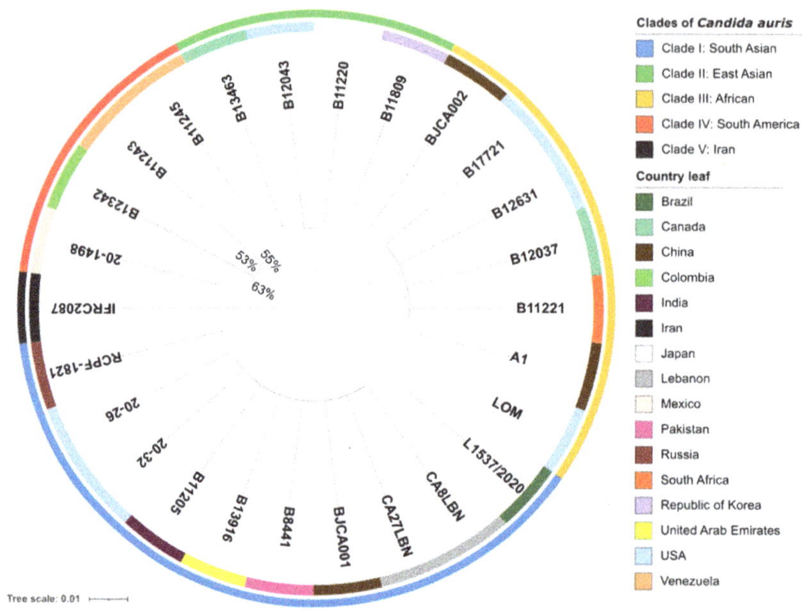

Figure 1. Phylogenomic tree of 26 strains of *Candida auris*, constructed using OrthoFinder with an orthologous gene model by performing 1000 bootstrap replicates. The 141,635 genes evaluated form a total of 5501 orthogroups, of which 4723 consist entirely of single-copy orthogroup genes existing in all the species. The outermost perimeter illustrates the geographic region corresponding to each of the five clades: clade I (blue), clade II (green), clade III (yellow), clade IV (red), and clade V (black). The labels of the leaves indicate the ID of each strain, and the color of each leaf refers to the country of origin of the strain.

The pan-genome analysis, conducted on the OrthoVenn3 server, created an orthologous clustering of the predicted proteins of the *C. auris* clades. *C. auris* IFRC2087 has a smaller proteome than the other isolates (*C. auris* B8441, *C. auris* B11220, *C. auris* B11221, *C. auris* B11243, and *C. auris* 20–1498). Based on 32,512 proteins, the OrthoVenn3 clustering displayed 5330 clusters, 4652 of which were single-copy clusters. Of the total number (in all 5 clades), those with at least 1 representative isolate of *C. auris* constituted 4881 core clusters (the core genome) (Figure 2A,B). A more limited comparison was performed between *C. auris* 20–1498 and three *C. auris* isolates from clade IV: *C. auris* B11243, *C. auris* B11245, and *C. auris* B12342. The evaluation of 21,925 proteins afforded 5271 clusters, of which 4749 were single-copy clusters and 4903 were core clusters (Figure 2C,D).

An evaluation was performed of the relation between the core genome of *C. auris* 20–1498 and two different groups: (1) all of the strains of the five clades, and (2) three strains in clade IV. At least 22 more clusters were observed when comparing the core genome of *C. auris* 20–1498 with the second versus the first group. The strain most closely linked to *C. auris* 20–1498 was *C. auris* B12342 from clade IV. According to the similarity matrix, the genomes of these two strains formed a great number of clusters (5222). The pairwise heatmaps of the number of overlapping clusters between each pair of *C. auris* species from the five different clades (Figure S1) and of the number of overlapping clusters between each pair of *C. auris* species from clade IV (Figure S2) are included in the Supplementary Material.

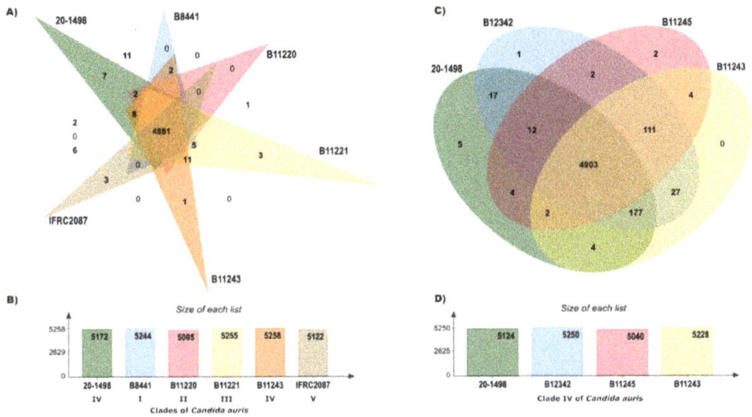

Figure 2. Summary of the pan-genome analysis of the orthologs carried out on the OrthoVenn3 server. (**A**) Venn diagram showing the distribution of orthologous clusters among the five clades of *C. auris*. (**B**) Summary of the proteins found in each strain of the five clades. (**C**) Venn diagram illustrating the distribution of orthologous clusters among the strains of *C. auris* in clade IV. (**D**) Summary of the proteins detected in each strain of clade IV.

3.3. Phylogenetic and Comparative Analysis of Erg11 and Fks1 Proteins

The phylogenetic tree of the Erg11 proteins from different yeasts consisted of 21 amino acid sequences of *Candida* spp., 1 of *Saccharomyces cerevisiae* S288C, and 1 of *Yarrowia lipolytica* CLIB122 as an outgroup. It was generated with the maximum likelihood method and the Le and Gascuel + G model (parameter = 0.9595) using MEGA7 software by performing 1000 bootstrap replicates (Figure 3A). On the other hand, the phylogenetic tree of the Fks1 proteins from different yeasts was comprised of 19 amino acid sequences of *Candida* spp., 1 of *Saccharomyces cerevisiae* S288C, and 1 of *Yarrowia lipolytica* CLIB122 as an outgroup. It was also generated with the maximum likelihood method and the Le and Gascuel + G model (parameter = 0.7018) using MEGA7 software by performing 1000 bootstrap replicates (Figure 3B).

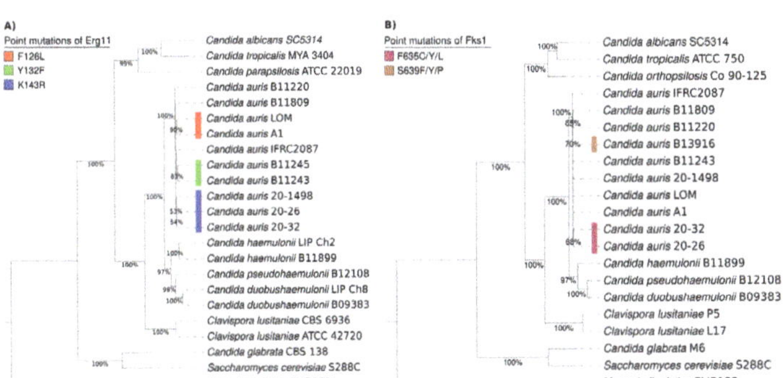

Figure 3. Phylogenetic trees of the Erg11 (**A**) and Fks1 proteins (**B**) and their putative orthologs in various *Candida* spp., 1 *Saccharomyces cerevisiae S288C* strain, and 1 *Yarrowia lipolytica CLIB122* strain as an outgroup. The point mutations of Erg11 and Fks1 from *C. auris* are illustrated by distinct colors on the phylogenetic trees. For the Erg11 point mutations, F126L is depicted in red, Y132F in green, and K143R in blue. For the Fks1 point mutations, F635C/Y/L is portrayed in purple and S639Y/P/F in brown.

Overall, the comparison of the amino acid sequences of the Erg11 and Fks1 proteins shows a close relationship between *C. auris*, *C. haemulonii*, *C. duobushaemulonii*, and *C. pseudohaemulonii* and justifies their grouping into a single clade. The present phylogenetic analysis confirmed that the Erg11 and Fks1 proteins from *C. haemulonii*, *C. duobushaemulonii*, and *C. pseudohaemulonii* are closely related to the same proteins in *C. auris* [26]. Moreover, the genotypes were determined for specific mutations in the Erg11 proteins (Y132F, K143R, and F126L) of *C. auris* strains associated with resistance to azoles. Based on the results, the K143R mutation detected in the lanosterol 14-alpha-demethylase (Erg11) of the *C. auris* 20–1498 isolate is probably related to the clinically observed resistance of this strain to fluconazole [10]. The K143R mutation has been predominately identified in clade I and in a few isolates from clade IV [1,3,27,28].

Regarding the S639Y/P/F and F635C/Y/L mutations in 1,3-beta-D-glucan synthase (Fks1), the most frequent mutation is S639P in isolates from clade IV. Likewise, S639F and S639Y are the most commonly identified mutations in micafungin-resistant isolates from clades I and III [3,27]. There are also reports of a F635C/Y/L mutation in isolates from clade I [28,29]. The fact that none of these mutations occurred in the *C. auris* 20–1498 strain is in accordance with the clinically observed susceptibility of the yeast to caspofungin [10].

3.4. Modeling the Candida auris 20–1498 Erg11 Protein, and Its Use for the Molecular Docking of Some Azoles

The structure of the Erg11 protein was built in 3D and then verified (Figure 4). The structure generated overlap with the template, indicating a high percentage of identity (72.78%) with it (Figure 4A). The Ramachandran plots constructed for the *Candida auris* 20–1498 Erg11 protein showed 89.6% of the residues located in favorable regions, demonstrating the reliability of the structure.

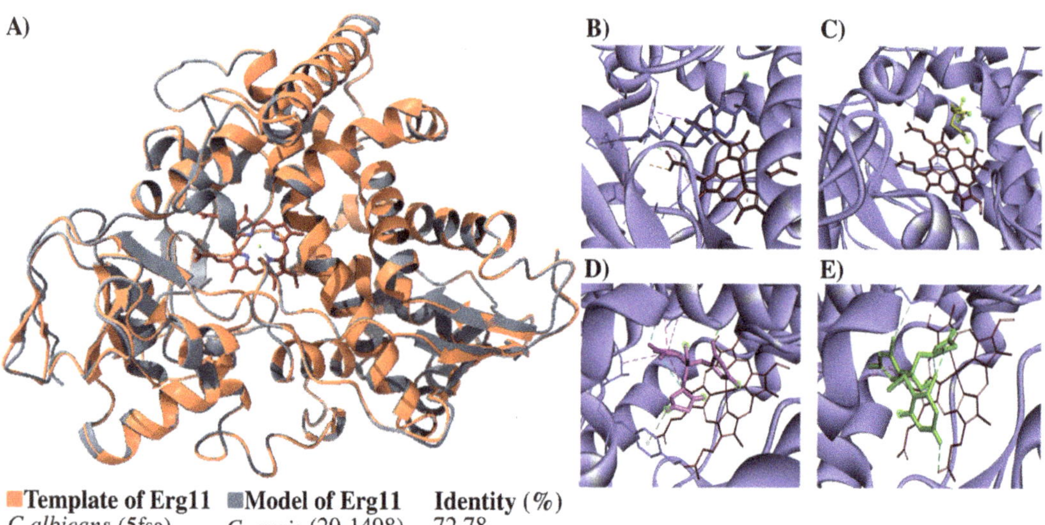

Figure 4. Modeling the *C. auris* 20–1498 Erg11 protein and its use for the molecular docking of some azoles. (**A**) The schematic illustration portrays the binding mode of the ligand with the Erg11 protein. The flat ribbon representation of the Erg11 of *Candida* spp. reveals the overlapping of *C. albicans* Erg11 (PDB: 5fsa) (orange) with *C. auris* 20–1498 Erg11 (gray). The heme as the prosthetic group is depicted in red (stick representation). The percentage of identity with their respective template is listed for the model. The predicted binding mode on *C. auris* 20–1498 Erg11 is shown for lanosterol (blue) (**B**), mevalonate (yellow) (**C**), fluconazole (pink) (**D**), and voriconazole (green) (**E**).

The next step was a coupling analysis to test the hypothesis that the protein had a higher affinity for the natural substrate lanosterol than for mevalonate (the negative control) and antifungals belonging to the azole family (Figure 4B–E). The docking study of the binding mode between lanosterol and the Erg11 protein (Table 2) evidenced a low binding energy value (high affinity). Meanwhile, there were higher binding energy values (lower affinity) for fluconazole and voriconazole on the Erg11 protein. The binding energies between the different ligands and the Erg11 protein K143 (wild-type) and R143 (substitution mutation) were also determined with a coupling analysis (Table 2) (Figure S3).

Table 2. Docking results of the binding mode between lanosterol, mevalonate, fluconazole, and voriconazole at the catalytic site of the *C. auris* 20–1498 Erg11 protein.

Molecules	Binding Energy (kcal/mol)	Interacting Residues	Interactions
Lanosterol	K143; R143 WT; M −12; −10.4	Val A: 304, HEM A: 525, Gly A: 307, Leu A: 376, Met A: 504, Pro A: 230, Leu A: 121, Phe A: 380, His A: 377, Phe A: 233, Ser A: 378, Tyr A: 118, Thr A: 122, Ile A: 131, Gly A: 303, Phe A: 126, Leu A: 300	Van Der Waals Pi-sigma Alkyl Pi-alkyl
Mevalonate	−5–5; −4.4	Gly A: 308, Phe A: 126, Tyr A: 132, Ile A: 131, Val A: 304, HEM A: 525, Leu A: 300, Gly A: 303, Gly A: 307	Van Der Waals Carbon hydrogen bond
Fluconazole	−8.8; −8.6	Leu A: 300, Val A: 304, Gly A: 303, Ile A: 131, Tyr A: 132, HEM A: 525, Phe A: 228, Tyr A: 118, Leu A: 376, Pro A: 375, Thr A: 311, His A: 310, Gly A: 307, Phe A: 126	Van Der Waals Conventional hydrogen bond Halogen (fluorine) Pi-donor hydrogen bond Amide-pi stacked Pi-alkyl
Voriconazole	−9.5; −7.3	Leu A: 376, Ile A: 379, Arg A: 381, Thr A: 311, Gly A: 308, Ile A: 131, Leu A: 300, Val A: 304, Gly A: 303, Phe A: 126, Thr A:122, Gly A: 307, Tyr A: 132, HEM A: 525	Van Der Waals Conventional hydrogen bond Carbon hydrogen bond Alkyl Pi-alkyl

The binding energy is expressed as kCal/mol (ΔG). Lanosterol, the natural substrate of the Erg11 enzyme, served as the positive binding control and mevalonate as the negative binding control. WT, wild-type; M, substitution mutation.

The models of the wild-type (K143) and mutant (K143R) Erg11 protein of *C. auris* 20–1498 are illustrated in Figure S3 (Supplementary Material). The docking results of the binding mode of lanosterol, mevalonate, fluconazole, and voriconazole at the catalytic site of each of these Erg11 proteins are shown in Table S2 (Supplementary Material).

3.5. Phenotypic Characteristics: Thermotolerance and Halotolerance

Thermotolerance and halotolerance have been described as characteristics that could help *C. auris* to survive in hospital environments. These characteristics may also help differentiate strains from the Metschnikowiaceae family, specifically between the *C. auris* clade and the *C. haemulonii* complex [11]. *C. auris* 20–1498 was found to be thermoresistant when incubated at 42 °C, unlike *C. haemulonii* 87, which was thermosensitive when incubated in poor media (YPD and SDA) at 37 °C and tolerant in rich media (e.g., BHI agar and blood agar) at 37 °C (Figure 5). Unlike *C. albicans* and *C. glabrata*, *C. auris* 20–1498 showed halotolerance, as it could resist NaCl at concentrations of 1 and 2 M.

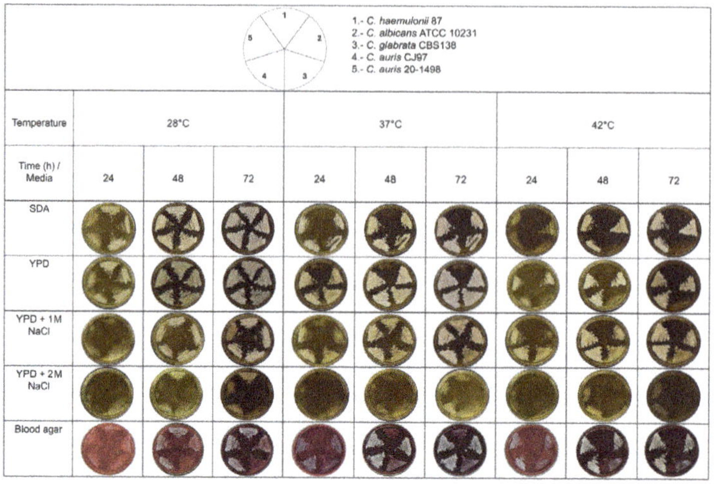

Figure 5. Thermotolerance and halotolerance phenotype of the Mexican *C. auris* 20–1498 compared to *C. auris* CJ97, *C. albicans* ATCC 10231, *C. glabrata* CBS 138, and *C. haemulonii* 87. The yeasts were grown in YPD broth under constant shaking at 28 °C until reaching the early stationary phase of growth (~15 h). The inoculum was adjusted to $As_{600} = 0.5$ with sterile YPD medium, and 5 µL of each strain was inoculated onto the corresponding culture media and streaked with a microbiological loop. The cultures were incubated at different temperatures (28, 37, and 42 °C), and the yeast growth was recorded every 24 h for 3 days. SDA, YPD, YPD-1 M NaCl, YPD-2 M NaCl, and blood agar served as the solid culture media [11].

4. Discussion

C. auris, an emerging fungal pathogen around the world, has been a challenge for major hospitals around the world because of its resistance to multiple antifungal agents. As a consequence, the treatment options are severely limited [30].

Before the first isolate of *C. auris* was detected in Mexico in 2020 [10], the strain was absent from national epidemiological and etiological reports on candidiasis [31]. The current contribution is the first attempt to define the molecular features of the *C. auris* isolate and carry out a phylogenomic search for mutations linked to its antifungal resistance. The short-read sequencing technique was utilized to produce a complete genome sequence of this pathogenic strain.

Whole-genome sequencing demonstrated the close relationship between the *C. auris* 20–1498 isolate and three strains of *C. auris* herein used to represent clade IV (containing strains endemic to South America). *C. auris* 20–1498 is more closely related to *C. auris* B12342 from Colombia than to *C. auris* B11243 and B11245 from Venezuela. The data here generated by whole-genome sequencing will serve to explore the population structure of *C. auris* 20–1498 and gain further insights into why certain strains are responsible for the multidrug resistance of a given clade. Such information should facilitate monitoring of the global dissemination of drug-resistant strains [32].

The pan-genomic analysis of *C. auris* strains in clade IV revealed an exclusive genome of the *C. auris* 20–1490 strain with five clusters. Each cluster contains different genes that encode proteins distinct from those encoded by other clusters. Thus, each cluster likely has a distinct function, defined by the assigned Gene Ontology (GO) term and Swiss-Prot Hit. Cluster 1 contains two proteins (with accession numbers 000009700.1 and 000309600.1) assigned the term GO:0005351 (F:carbohydrate:proton symporter activity) and the Swiss-Prot Hit code A0A1D8PCL1 (high-affinity glucose transporter). Cluster 2 also contains two proteins (with accession numbers 000076100.1 and 000194600.1) given the term GO:0055085 (P:transmembrane transport; IDA:SGD) and the Swiss-Prot Hit code P13587 (sodium transport ATPase 1). Cluster 3 contains two proteins (with accession numbers 000353300.1 and 000370500.1) designated by the term GO:0005524 (F: ATP binding) and the Swiss-Prot Hit code: P53623 (heat shock protein 70). Finally, clusters 4 and 5 each had two hypothetical proteins: the former 000305000.1 and 000326000.1, and the latter 000119800.1 and 000521300.1. These were not furnished a GO term or Swiss-Prot Hit code.

Because mutations vary in relation to the clade and country of origin of a species, mutations are examined in relation to the antifungal resistance of each specific clade. Based on the minimum inhibitory concentration (MIC) breakpoints recommended by the CDC and the Clinical Laboratory Standard Institute (CLSI, supplement M60) [33], *C. auris* 20–1498 is resistant to fluconazole ($\geq$64 μg/mL) and amphotericin B ($\geq$2 μg/mL) but susceptible to caspofungin ($\geq$0.5 μg/mL) [10]. When the phenotypic test was matched with the genotypic results of the *C. auris* 20–1498 isolate, it was not surprising to find the K143R mutation, which has been reported in a few fluconazole-resistant isolates of *C. auris* belonging to clade IV. This substitution is linked to the elevated MICs of azoles [3,34]. According to the docking study, the Erg11 protein has greater affinity for its substrate (lanosterol) than for two of the main antifungals belonging to the azole group (fluconazole and voriconazole).

Two other mutations in Fsk1 (S639F and F635) are linked to multidrug resistance in various *C. auris* strains. Neither one was found in *C. auris* 20–1498. S639F has been detected in the Fks1 of the multidrug-resistant *C. auris* B13916 strain, while F635C has been identified in the pandrug-resistant *C. auris* 20–26 and *C. auris* 20–32 strains. For the latter strains, the MIC of echinocandin is high [28]. The docking simulations with the Fks1 protein are in agreement with the results concerning the caspofungin sensitivity. The Mexican patient evolved favorably with systemic antifungal therapy with caspofungin [10].

Studies on evolved strains of *C. auris* have revealed multiple novel mechanisms of multidrug resistance [35]. With mutations in *ERG3* and *CIS2*, there is a mutation in the transcription factor TAC1b and an overexpression of the drug efflux pump Cdr1, leading to a higher MIC for echinocandin. Some of the mechanisms of resistance to amphotericin B are known to be related to the expression levels of genes in the ergosterol biosynthesis pathway. Evaluation of the reverse transcription PCR results demonstrated that Upc2 regulates *ERG11* expression and also activates the Mrr1/Mdr1 pathway [36]. In the current contribution, the point mutation V704L in Cdr1 was not detected in the genome of *C. auris* 20–1498. Future research can take advantage of the *C. auris* 20–1498 genome database herein generated in order to carry out gene expression studies with the aim of exploring new resistance mechanisms in this strain and other clade IV isolates.

The present analysis of the resistance mechanisms of *C. auris* focused on mutations in the *ERG11* and *FKS1* genes. Besides being the genes associated with resistance to azoles and echinocandins, which together with amphotericin B constitute the drugs of choice for antifungal treatment in Mexico, they are the markers present in the greatest abundance in the current gene databases and have been reported in practically all the clades of *C. auris* (I-V). Hence, research on these markers would allow for epidemiological and comparative analysis to be carried out, even without having sequenced the complete genome.

In Mexico, 20–1498 is the first known strain of *C. auris* isolated from a patient, and no further information exists in this country on mutations in the Erg11 and Fks1 proteins in *C. auris*. It is very important to evaluate the latter proteins in relation to the antifungals recommended for the treatment of invasive candidiasis in Mexico, with fluconazole being the first drug of choice, followed by echinocandins, voriconazole, and amphotericin B [10]. The resulting information on the point mutations is important for the establishment of accurate antifungal resistance and antifungal susceptibility testing in healthcare settings. Such testing would be invaluable in determining appropriate therapeutic strategies. On the other hand, since the genome sequence of the *C. auris* 20–1498 strain has been deposited in the NCBI database, it can provide a model for further research on resistance, virulence factors, molecular epidemiology, therapeutic targets, and antifungal design.

According to the thermotolerance and halotolerance capacity found for the *C. auris* 20–1498 strain, it likely emerged from a natural reservoir, a conclusion supported by genomic evidence and the ecology of related fungal species [37]. Since pathogenic *C. auris* can tolerate high concentrations of salt, it likely evolved in niches of marine ecosystems [37]. Its other potential environmental sources include terrestrial and freshwater reservoirs, with specific niches in soil, plants, and animals.

The characteristics of *C. auris* in relation to thermotolerance (at 40–42 °C) and halotolerance to NaCl have been analyzed [37,38]. It is suggested that thermotolerance and halotolerance would be advantageous for the survival of fungi on the skin, axilla, and groin, the most common sites of *C. auris* isolation in intra-hospital environments [37]. *C. auris* 20–1498 was isolated from an in-hospital environment, a hospital that months later was converted into a COVID-19 unit [39]. Therefore, this hospital should be monitored for the persistence of the same strain.

The current data can be used to study fungal biology and virulence in order to provide greater insight into the phylogenetic relationships between multidrug-resistant *C. auris* strains and to determine which genomic regions are associated with specific phenotypes. The information on the genome of *C. auris* will allow for further related research, such as comparative analysis and the evolution of the genomes, on the first *C. auris* strain isolated in Mexico and the one isolated months later in the same hospital in a patient with COVID-19 [37,38]. The focus on whole-genome sequencing and the identification of mutations linked to the drug resistance of fungi could lead to the discovery of new therapeutic targets as well as new antifungals capable of responding to the serious problem of the multidrug resistance of *C. auris*.

5. Conclusions

The first isolate of *C. auris* detected in Mexico is the 20–1498 strain. It has a total genome size of 12.86 Mbp and an average guanine-cytosine (GC) content of 45.5%. Genome annotation revealed a total of 5432 genes encoding 5515 proteins. The genomic analysis demonstrated that the *C. auris* 20–1498 strain belongs to clade IV (containing strains endemic to South America). Of the two genes (*ERG11* and *FKS1*) associated with drug resistance in *C. auris*, a mutation was found in the K143R gene located in a mutation hotspot of *ERG11* (lanosterol 14-α-demethylase), an antifungal drug target of azoles. The Cdr1 point mutation V704L was not detected in the genome of *C. auris* 20–1498. The current results can be used to study the fungal biology and virulence in order to provide greater insight into the phylogenetic relationships between multidrug-resistant *C. auris* strains and to determine which genomic regions are associated with specific phenotypes. The focus on whole-genome sequencing and the identification of mutations linked to the drug resistance of fungi could lead to the discovery of new therapeutic targets and new antifungal compounds.

Supplementary Materials: The following supporting information can be downloaded at: https://www.mdpi.com/article/10.3390/jof10060392/s1.

Author Contributions: Conceptualization, L.V.-T. and C.H.-R.; writing—original draft preparation, A.C.-R. and L.V.-T.; writing—review and editing, A.C.-R., L.V.-T., and C.H.-R.; genome assembly and annotation, A.C.-R. and J.A.H.-G.; whole-genome sequencing analysis, A.C.-R. and J.A.H.-G.; phylogenomic and pan-genomic analysis, A.C.-R. and C.B.-C.; phylogenetic and comparative analysis of Erg11 and Fks1, A.C.-R.; clinical studies of the isolate and identification by sequencing of the *ITS* marker G.M.G.; phenotypic tests, A.V.-M.; project administration, L.V.-T. and C.H.-R. All authors have read and agreed to the published version of the manuscript.

Funding: Instituto Politécnico Nacional (SIP20240946, SIP20231481, SIP20231480, SIP20220742, SIP20220795); CONACyT 283225.

Institutional Review Board Statement: Not applicable.

Informed Consent Statement: Not applicable.

Data Availability Statement: The *Candida auris* 20–1498 genome sequence herein generated was deposited in the GenBank BioProject (PRJNA1013603) with genome annotation and BioSample ID (SAMN39051800).

Acknowledgments: We thank Bernard Dujon of the Pasteur Institute for the donation of the *C. glabrata* CBS 138 strain; and Eulogio Valentín Gómez from the University of Valencia, Spain, and Javier Pemán García from the Hospital Universitario y Politécnico La Fe, Valencia, Spain for the donation of the *C. auris* CJ97 strain.

Conflicts of Interest: The authors declare no conflicts of interest.

References

1. Satoh, K.; Makimura, K.; Hasumi, Y.; Nishiyama, Y.; Uchida, K.; Yamaguchi, H. *Candida auris* sp. nov., a novel ascomycetous yeast isolated from the external ear canal of an inpatient in a Japanese hospital. *Microbiol. Immunol.* **2009**, *53*, 41–44. [CrossRef] [PubMed]
2. Forsberg, K.; Woodworth, K.; Walters, M.; Berkow, E.L.; Jackson, B.; Chiller, T.; Vallabhaneni, S. *Candida auris*: The recent emergence of a multidrug-resistant fungal pathogen. *Med. Mycol.* **2019**, *57*, 1–12. [CrossRef] [PubMed]
3. Chow, N.A.; Muñoz, J.F.; Gade, L.; Berkow, E.L.; Li, X.; Welsh, R.M.; Forsberg, K.; Lockhart, S.R.; Adam, R.; Alanio, A.; et al. Tracing the Evolutionary History and Global Expansion of *Candida auris* Using Population Genomic Analyses. *mBio* **2020**, *11*, e03364-19. [CrossRef] [PubMed]
4. Lockhart, S.R.; Etienne, K.A.; Vallabhaneni, S.; Farooqi, J.; Chowdhary, A.; Govender, N.P.; Colombo, A.L.; Calvo, B.; Cuomo, C.A.; Desjardins, C.A.; et al. Simultaneous Emergence of Multidrug-Resistant *Candida auris* on 3 Continents Confirmed by Whole-Genome Sequencing and Epidemiological Analyses. *Clin. Infect. Dis.* **2017**, *64*, 134–140. [CrossRef] [PubMed]
5. Chow, N.A.; de Groot, T.; Badali, H.; Abastabar, M.; Chiller, T.M.; Meis, J.F. Potential Fifth Clade of *Candida auris*, Iran, 2018. *Emerg. Infect. Dis.* **2019**, *25*, 1780–1781. [CrossRef] [PubMed]
6. Spruijtenburg, B.; Badali, H.; Abastabar, M.; Mirhendi, H.; Khodavaisy, S.; Sharifisooraki, J.; Taghizadeh Armaki, M.; de Groot, T.; Meis, J.F. Confirmation of fifth *Candida auris* clade by whole genome sequencing. *Emerg. Microbes Infect.* **2022**, *11*, 2405–2411. [CrossRef] [PubMed]
7. Suphavilai, C.; Ko, K.K.K.; Lim, K.M.; Tan, M.G.; Boonsimma, P.; Keat Chu, J.J.; Goh, S.S.; Rajandran, P.; Nagarajan, N. Discovery of the Sixth *Candida auris* Clade in Singapore.medRxiv. *medRxiv.* **2023**. [CrossRef]
8. Muñoz, J.F.; Welsh, R.M.; Shea, T.; Batra, D.; Gade, L.; Howard, D.; Rowe, L.A.; Meis, J.F.; Litvintseva, A.P.; Cuomo, C.A. Clade-specific chromosomal rearrangements and loss of subtelomeric adhesins in *Candida auris*. *Genetics* **2021**, *218*, iyab029. [CrossRef]
9. Chowdhary, A.; Prakash, A.; Sharma, C.; Kordalewska, M.; Kumar, A.; Sarma, S.; Tarai, B.; Singh, A.; Upadhyaya, G.; Upadhyay, S.; et al. A multicentre study of antifungal susceptibility patterns among 350 *Candida auris* isolates (2009-17) in India: Role of the ERG11 and FKS1 genes in azole and echinocandin resistance. *J. Antimicrob. Chemother.* **2018**, *73*, 891–899. [CrossRef]
10. Ayala-Gaytán, J.J.; Montoya, A.M.; Martínez-Resendez, M.F.; Guajardo-Lara, C.E.; de JTreviño-Rangel, R.; Salazar-Cavazos, L.; Llaca-Díaz, J.M.; González, G.M. First case of *Candida auris* isolated from the bloodstream of a Mexican patient with serious gastrointestinal complications from severe endometriosis. *Infection* **2021**, *49*, 523–525. [CrossRef]
11. Reséndiz-Sánchez, J.; Ortiz-Álvarez, J.; Casimiro-Ramos, A.; Hernández-Rodríguez, C.; Villa-Tanaca, L. First report of a catheter-related bloodstream infection by *Candida haemulonii* in a children's hospital in Mexico City. *Int. J. Infect. Dis.* **2020**, *92*, 123–126. [CrossRef] [PubMed]
12. Schmieder, R.; Edwards, R. Quality control and preprocessing of metagenomic datasets. *Bioinformatics* **2011**, *27*, 863–864. [CrossRef] [PubMed]
13. Bolger, A.M.; Lohse, M.; Usadel, B. Trimmomatic: A flexible trimmer for Illumina sequence data. *Bioinformatics* **2014**, *30*, 2114–2120. [CrossRef] [PubMed]
14. Zerbino, D.R.; Birney, E. Velvet: Algorithms for de novo short read assembly using de Bruijn graphs. *Genome Res.* **2008**, *18*, 821–829. [CrossRef]
15. Gurevich, A.; Saveliev, V.; Vyahhi, N.; Tesler, G. QUAST: Quality assessment tool for genome assemblies. *Bioinformatics* **2013**, *29*, 1072–1075. [CrossRef] [PubMed]
16. Steinbiss, S.; Silva-Franco, F.; Brunk, B.; Foth, B.; Hertz-Fowler, C.; Berriman, M.; Otto, T.D. Companion: A web server for annotation and analysis of parasite genomes. *Nucleic Acids Res.* **2016**, *44*, W29–W34. [CrossRef]
17. Sayers, E.W.; Bolton, E.E.; Brister, J.R.; Canese, K.; Chan, J.; Comeau, D.C.; Connor, R.; Funk, K.; Kelly, C.; Kim, S.; et al. Database resources of the National Center for Biotechnology Information. *Nucleic Acids Res.* **2022**, *50*, D20–D26. [CrossRef]
18. Emms, D.M.; Kelly, S. OrthoFinder: Phylogenetic orthology inference for comparative genomics. *Genome Biol.* **2019**, *20*, 238. [CrossRef]
19. Sun, J.; Lu, F.; Luo, Y.; Bie, L.; Xu, L.; Wang, Y. OrthoVenn3: An integrated platform for exploring and visualizing orthologous data across genomes. *Nucleic Acids Res.* **2023**, *51*, W397–W403. [CrossRef]
20. Letunic, I.; Bork, P. Interactive Tree Of Life (iTOL) v5: An online tool for phylogenetic tree display and annotation. *Nucleic Acids Res.* **2021**, *49*, W293–W296. [CrossRef]

21. Kumar, S.; Stecher, G.; Tamura, K. MEGA7: Molecular Evolutionary Genetics Analysis Version 7.0 for Bigger Datasets. *Mol. Biol. Evol.* **2016**, *33*, 1870–1874. [CrossRef] [PubMed]
22. Hargrove, T.Y.; Friggeri, L.; Wawrzak, Z.; Qi, A.; Hoekstra, W.J.; Schotzinger, R.J.; York, J.D.; Guengerich, F.P.; Lepesheva, G.I. Structural analyses of *Candida albicans* sterol 14α-demethylase complexed with azole drugs address the molecular basis of azole-mediated inhibition of fungal sterol biosynthesis. *J. Biol. Chem.* **2017**, *292*, 6728–6743. [CrossRef] [PubMed]
23. Eswar, N.; Webb, B.; Marti-Renom, M.A.; Madhusudhan, M.S.; Eramian, D.; Shen, M.Y.; Pieper, U.; Sali, A. Comparative protein structure modeling using Modeller. *Curr. Protoc. Bioinform.* **2006**, *54*, 5–6. [CrossRef]
24. Gómez-García, O.; Andrade-Pavón, D.; Campos-Aldrete, E.; Ballinas-Indilí, R.; Méndez-Tenorio, A.; Villa-Tanaca, L.; Álvarez-Toledano, C. Synthesis, Molecular Docking, and Antimycotic Evaluation of Some 3-Acyl Imidazo [1,2-a] pyrimidines. *Molecules* **2018**, *23*, 599. [CrossRef] [PubMed]
25. Eberhardt, J.; Santos-Martins, D.; Tillack, A.F.; Forli, S. AutoDock Vina 1.2.0: New Docking Methods, Expanded Force Field, and Python Bindings. *J. Chem. Inf. Model.* **2021**, *61*, 3891–3898. [CrossRef]
26. Rodrigues, L.S.; Gazara, R.K.; Passarelli-Araujo, H.; Valengo, A.E.; Pontes PV, M.; Nunes-da-Fonseca, R.; de Souza, R.F.; Venancio, T.M.; Dalla-Costa, L.M. First Genome Sequences of Two Multidrug-Resistant *Candida haemulonii* var. vulnera Isolates From Pediatric Patients With Candidemia. *Front. Microbiol.* **2020**, *11*, 1535. [CrossRef] [PubMed]
27. Li, D.; Wang, Y.; Hu, W.; Chen, F.; Zhao, J.; Chen, X.; Han, L. Application of Machine Learning Classifier to *Candida auris* Drug Resistance Analysis. *Front. Cell. Infect. Microbiol.* **2021**, *11*, 742062. [CrossRef] [PubMed]
28. Jacobs, S.E.; Jacobs, J.L.; Dennis, E.K.; Taimur, S.; Rana, M.; Patel, D.; Gitman, M.; Patel, G.; Schaefer, S.; Iyer, K.; et al. *Candida auris* Pan-Drug-Resistant to Four Classes of Antifungal Agents. *Antimicrob. Agents Chemother.* **2022**, *66*, e0005322. [CrossRef] [PubMed]
29. Kiyohara, M.; Miyazaki, T.; Okamoto, M.; Hirayama, T.; Makimura, K.; Chibana, H.; Nakada, N.; Ito, Y.; Sumiyoshi, M.; Ashizawa, N.; et al. Evaluation of a Novel FKS1 R1354H Mutation Associated with Caspofungin Resistance in *Candida auris* Using the CRISPR-Cas9 System. *J. Fungi* **2023**, *9*, 529. [CrossRef]
30. Reslan, L.; Araj, G.F.; Finianos, M.; El Asmar, R.; Hrabak, J.; Dbaibo, G.; Bitar, I. Molecular Characterization of *Candida auris* Isolates at a Major Tertiary Care Center in Lebanon. *Front. Microbiol.* **2022**, *12*, 770635. [CrossRef]
31. Reyes-Montes MD, R.; Duarte-Escalante, E.; Martínez-Herrera, E.; Acosta-Altamirano, G.; Frías-De León, M.G. Current status of the etiology of candidiasis in Mexico. *Rev. Iberoam. Micol.* **2017**, *34*, 203–210. [CrossRef]
32. Muñoz, J.F.; Gade, L.; Chow, N.A.; Loparev, V.N.; Juieng, P.; Berkow, E.L.; Farrer, R.A.; Litvintseva, A.P.; Cuomo, C.A. Genomic insights into multidrug-resistance, mating and virulence in *Candida auris* and related emerging species. *Nat. Commun.* **2018**, *9*, 5346. [CrossRef] [PubMed]
33. Centers for Disease Control and Prevention. Antifungal Susceptibility Testing and Interpretation | Candida auris | Fungal Diseases | CDC. 2023. Available online: https://www.cdc.gov/candida-auris/hcp/laboratories/antifungal-susceptibility-testing.html (accessed on 28 May 2024).
34. Healey, K.R.; Kordalewska, M.; Jiménez Ortigosa, C.; Singh, A.; Berrío, I.; Chowdhary, A.; Perlin, D.S. Limited ERG11 Mutations Identified in Isolates of *Candida auris* Directly Contribute to Reduced Azole Susceptibility. *Antimicrob. Agents Chemother.* **2018**, *62*, e01427-18. [CrossRef] [PubMed]
35. Carolus, H.; Pierson, S.; Muñoz, J.F.; Subotić, A.; Cruz, R.B.; Cuomo, C.A.; Van Dijck, P. Genome-Wide Analysis of Experimentally Evolved *Candida auris* Reveals Multiple Novel Mechanisms of Multidrug Resistance. *mBio* **2021**, *12*, e03333-20. [CrossRef]
36. Li, J.; Aubry, L.; Brandalise, D.; Coste, A.T.; Sanglard, D.; Lamoth, F. Upc2-mediated mechanisms of azole resistance in *Candida auris*. *Microbiol. Spectr.* **2024**, *12*, e0352623. [CrossRef] [PubMed]
37. Akinbobola, A.B.; Kean, R.; Hanifi, S.M.A.; Quilliam, R.S. Environmental reservoirs of the drug-resistant pathogenic yeast *Candida auris*. *PLoS Pathog.* **2023**, *19*, e1011268. [CrossRef] [PubMed]
38. Jackson, B.R.; Chow, N.; Forsberg, K.; Litvintseva, A.P.; Lockhart, S.R.; Welsh, R.; Vallabhaneni, S.; Chiller, T. On the Origins of a Species: What Might Explain the Rise of *Candida auris*? *J. Fungi* **2019**, *5*, 58. [CrossRef]
39. Villanueva-Lozano, H.; Treviño-Rangel, R.J.; González, G.M.; Ramírez-Elizondo, M.T.; Lara-Medrano, R.; AlemanBocanegra, M.C.; Guajardo-Lara, C.E.; Gaona-Chávez, N.; Castilleja-Leal, F.; Torre-Amione, G.; et al. Outbreak of *Candida auris* infection in a COVID-19 hospital in Mexico. *Clin. Microbiol. Infect.* **2021**, *27*, 813–816. [CrossRef]

Journal of
Fungi

MDPI

Article

Molecular Evaluation of the mRNA Expression of the *ERG11*, *ERG3*, *CgCDR1*, and *CgSNQ2* Genes Linked to Fluconazole Resistance in *Candida glabrata* in a Colombian Population

Leidy Yurany Cárdenas Parra [1,2], Ana Elisa Rojas Rodríguez [2,*], Jorge Enrique Pérez Cárdenas [1] and Juan Manuel Pérez-Agudelo [1]

1 Facultad de Ciencias para la Salud, Universidad de Caldas, Manizales 170004, Colombia; leidy.cardenas@ucaldas.edu.co (L.Y.C.P.); labmicro@ucaldas.edu.co (J.E.P.C.); juan.perez@ucaldas.edu.co (J.M.P.-A.)

2 Facultad de Ciencias de la Salud, Universidad Católica de Manizales, Manizales 170001, Colombia

* Correspondence: arojas@ucm.edu.co; Tel.: +57-6068933050 (ext. 3142)

Abstract: Introduction: The study of *Candida glabrata* genes associated with fluconazole resistance, from a molecular perspective, increases the understanding of the phenomenon with a view to its clinical applicability. Objective: We sought to establish the predictive molecular profile of fluconazole resistance in *Candida glabrata* by analyzing the *ERG11*, *ERG3*, *CgCDR1*, and *CgSNQ2* genes. Method: Expression was quantified using RT-qPCR. Metrics were obtained through molecular docking and Fisher discriminant functions. Additionally, a predictive classification was made against the susceptibility of *C. glabrata* to fluconazole. Results: The relative expression of the *ERG3*, *CgCDR1*, and *CgSNQ2* genes was higher in the fluconazole-resistant strains than in the fluconazole-susceptible, dose-dependent strains. The gene with the highest relative expression in the fluconazole-exposed strains was *CgCDR1*, and in both the resistant and susceptible, dose-dependent strains exposed to fluconazole, this was also the case. The molecular docking model generated a median number of contacts between fluconazole and *ERG11* that was lower than the median number of contacts between fluconazole and *ERG3*, -*CgCDR1*, and -*CgSNQ2*. The predicted classification through the multivariate model for fluconazole susceptibility achieved an accuracy of 73.5%. Conclusion: The resistant strains had significant expression levels of genes encoding efflux pumps and the *ERG3* gene. Molecular analysis makes the identification of a low affinity between fluconazole and its pharmacological target possible, which may explain the lower intrinsic susceptibility of the fungus to fluconazole.

Keywords: *Candida glabrata*; fungal drug resistance; reverse transcriptase polymerase chain reaction; fluconazole; molecular docking simulation; multivariate analysis

1. Introduction

Candida spp. is an etiological agent frequently isolated in infections, particularly, those caused by *Candida glabrata*, which are crucial, and so evidence must be collected to increase the understanding of the molecular implications, as well as an appropriate therapeutic approach, given the increasing incidence of such infections, especially in immunocompromised patients and its ability to develop resistance to or its susceptibility to being dose-dependent on a group of commonly used antifungals [1–4]. Studying the molecular mechanisms underlying drug resistance is essential for designing more effective therapeutic strategies and for anticipating and properly managing infections by this yeast in clinical settings. The scientific literature has documented various genetic and epigenetic factors that contribute to azole resistance in *C. glabrata*, providing details that can be considered for various research purposes [1,2].

In Colombia, *C. glabrata* is a pathogen that significantly contributes to the burden of fungal diseases. Epidemiological studies have shown that the prevalence of *C. glabrata* is increasing, particularly in clinical and hospital settings [3–7]. The frequency of isolation and the resistance or dose-dependent susceptibility exhibited by *C. glabrata* to azoles represent a significant therapeutic challenge and underline the importance of studies aimed at improving therapeutic effectiveness, as well as the continuous surveillance of antifungal resistance in the country.

Resistance can be caused by multiple factors, for example, frequent exposure of the fungus to antifungals, where its ability to evolve and adapt to hostile environments ensures its survival. Likewise, the exposure of the biological agent to conventional drug action mechanisms and the lack of new antifungal active principles with alternative mechanisms of action make treatment complex due to resistance, which is due to the structural and functional biological adaptive adjustments of the fungi, resulting in decreased pharmacological effectiveness [8,9].

Typically, azoles act by blocking lanosterol 14-alpha-demethylase, a cytochrome P450 enzyme that converts lanosterol, a structural intermediate of the fungal membrane, into ergosterol. From a molecular perspective, the free nitrogen atom located in the azole ring of fluconazole binds to the iron atom of the heme group of lanosterol 14-alpha-demethylase, disrupting the enzyme's catalytic activity and, thus, interrupting the formation of the fungal membrane [10].

A resistance-associated mechanism involves the drug's extrusion from inside the fungal cell, in which proteins or pumps belonging to two superfamilies, ATP Binding Cassette and MFS, are involved, and they are encoded by the *CgCDR1*, *CgSNQ2*, *PDH1*, and *MDR* genes [11–21]. Another resistance mechanism is associated with alterations in expression and mutations of the genes encoding the enzymes involved in the ergosterol biosynthesis pathway, such as lanosterol 14-alpha-demethylase and C-5 desaturase, which are responsible for synthesizing toxic sterols that accumulate in a fungus when it is exposed to azoles.

The *ERG11* gene encodes the lanosterol 14-alpha-demethylase enzyme, and its overexpression and/or mutations are associated with resistance, and this biological basis responds to maintaining the integrity of the fungal membrane in the presence of azoles and decreasing the affinity between the azole and the enzyme acting as the pharmacological target. On the other hand, mutations in the *ERG3* gene, which encodes the C-5 desaturase enzyme, have been described as resistance-associated mechanisms, as the enzyme loses activity and stops synthesizing toxic sterols in the presence of azoles, facilitating the survival of a fungus. Moreover, pathways for the synthesis of constitutive compounds of the fungal membrane have been found to be activated [22–25]. These genes have been identified in previous studies as important determinants of azole resistance in *C. glabrata*, making them ideal for molecular research aimed at understanding its resistance mechanisms [11,23,26].

Overall, the described mechanisms allow this fungus to avoid or reduce the effectiveness of azoles, a situation that necessitates further research to improve the understanding of this phenomenon at a local level and from a molecular perspective to make the problem visible in order to advance a comprehensive approach. In this study, the expression of the *ERG3*, *ERG11*, *CgSNQ2*, and *CgCDR1* genes of *C. glabrata* was evaluated using reverse transcriptase polymerase chain reaction (RT-qPCR), and high-level computational and data analytics routes were involved in contributing to the understanding and explanation of the phenomenon.

2. Materials and Methods

2.1. Study Type

A general experimental approach was carried out, contributing to the explanatory level of the research.

2.2. Strains and Sample-Handling Conditions

Sixteen strains of *C. glabrata*, collected in a previous study conducted in an intensive care unit in the city of Manizales by the Biosalud research group at the Universidad de

Caldas, were used. Species confirmation of these strains was performed using the Vitek 2 compact system (Biomerieux, Marcy-l'Étoile, France), and their viability was established by adequate growth in culture media such as Sabouraud (Scharlau Microbiology, Sentmenat, Spain) and potato dextrose agar (PDA, Scharlau Microbiology, Sentmenat, Spain).

2.3. Antifungal Susceptibility Testing

Each strain underwent antifungal susceptibility testing in accordance with the protocols proposed by the Clinical and Laboratory Standards Institute (CLSI) version M27-A4 [27]. *Candida* isolates were grown on Sabouraud dextrose agar (Scharlau, Microbiology, Sentmenat, Spain) at 37 °C for 48 h. Antifungal susceptibility testing was performed in 96-well plates. Briefly, colonies of *C. glabrata* were re-suspended in RPMI 1640 medium (MP Biomedicals, LLC, Santa Ana, CA, USA), resulting in final concentrations of 0.5×10^3 to 2.5×10^3 cells/mL in the inoculum. Samples in the microplates were incubated at 37 °C for 48 h, and the minimum inhibitory concentrations (MICs) were evaluated spectrophotometrically at A 600 using a microplate reader. The antifungal drug tested in this study was fluconazole, which was dissolved and diluted according to CLSI recommendations [27]. MIC endpoints were assigned as the lowest concentration of antifungal drug that resulted in 50% growth inhibition compared with the drug-free control well for fluconazole. *Candida albicans* ATCC 90028 and *Candida glabrata* ATCC 90030 were used as the quality control organisms [27].

The MIC endpoint was used to interpret the resistance breakpoint values of the antifungal drugs according to CLSI guidelines and subsequent reports [28,29]. The MIC breakpoint values are as follows: resistant MIC > 64 µg/mL and SDD MIC ≤ 32 µg/mL.

2.4. Fluconazole Exposure Test

The fluconazole exposure assay was developed using liquid Sabouraud medium (Scharlau, Microbiology, Sentmenat, Spain) as the culture medium, and the established concentrations of the antifungal were used according to the protocol proposed by the CLSI broth dilution method. All tubes were incubated at 35 °C until the logarithmic phase of yeast growth occurred. The sediment size in each tube containing fluconazole (test group: exposed to fluconazole) was compared with the positive control (control group: not exposed to fluconazole). Cells were collected from both the positive control tube and the fluconazole tube, where cell growth was adjusted to a concentration between 2×10^8 and 3×10^8 yeast/mL. The antifungal concentration of the collected samples exposed to fluconazole is shown in Table 1.

Table 1. Fluconazole concentrations of the collected samples exposed to the antifungal.

Sample	Fluconazole Concentration
06-3A	8 µg/mL
95-1A	4 µg/mL
116-5	8 µg/mL
137-1	4 µg/mL
184-1A	8 µg/mL
65-3	16 µg/mL
11-3A	16 µg/mL
04-1A	32 µg/mL
16-5A	8 µg/mL
24-5A	8 µg/mL
25-1A	16 µg/mL
20-2A	16 µg/mL
14-5A	32 µg/mL
07-1A	8 µg/mL
68-4	8 µg/mL
109-1A	8 µg/mL

The tubes selected in the antifungal exposure test were centrifuged to obtain yeast sediment, which was stored in DNase- and RNase-free Eppendorf tubes, and RNAlater (Invitrogen™: Thermo Fisher Scientific, Waltham, MA, USA) was later added to preserve them at −80 °C until extraction.

2.5. RNA Extraction

In the gene expression experiments performed, the guidelines proposed in the Minimum Information for Publication of Quantitative Real-Time PCR Experiments (MIQE) were followed. For RNA extraction, the protocol suggested in the extraction kit used (RiboPure–Yeast Kit, Thermo Fisher Scientific, Waltham, MA, USA) was followed according to the manufacturer's recommendations.

2.6. Quality Criteria for RNA Extraction

The RNA solution was subjected to 1% agarose gel electrophoresis to observe the integrity of the genetic material. The purity of the RNA was measured using a UVIS Drop UVS99™ spectrophotometer (Thermo Fisher Scientific, Waltham, MA, USA) with readings of sample absorbance at 230, 260, and 280 nm. RNA purity values with a ratio of 260:280 and 260:230 greater than or equal to 2 were considered high purity and subsequently used for the RT-qPCR.

To evaluate the presence of DNA after treatment with DNAase I, PCR amplification of one of the RNA samples, without reverse transcription and in the presence of a universal primer for *C. glabrata*, was performed to assess if there was amplification of residual DNA. The DNA-contaminated strains underwent a new DNAase I treatment.

2.7. Protocol Standardization

For standardization, the *C. glabrata* ATCC 90030 strain was used as a reference for a susceptible dose-dependent strain, and a strain from the fluconazole-resistant group was randomly chosen. All procedures were performed according to previously established guidelines to ensure adequate reaction purity and to prevent RNA contamination by other microorganisms.

2.8. Reverse Transcription (RT) Protocol

The reverse transcription process was performed using a StepOnePlus Real-Time PCR System thermocycler (Applied Biosystems, Waltham, MA, USA) with the following protocol: incubation at 25 °C for 10 min and again at 42 °C for 15 min, and then the enzyme was inactivated at 85 °C for 5 min. The cDNA was converted from 1 μg of purified RNA using a SensiFastTM cDNA kit (Bioline, London, UK). The manufacturer's instructions were meticulously followed without making any modifications to the protocol.

2.9. Selection and Optimization of the Primers for the qPCR

The primers for the reaction, both for the endogenous reference control gene and the resistance genes (*ERG3*, *ERG11*, *CgSNQ2*, and *CgCDR1*), are shown in Table 2.

Table 2. Primer Sequences.

Gene		Primer Sequence (5′–3′)	Reference
CgCDR1	Forward	CATACAAGAAACACCAAAGTCGGT	[30]
	Reverse	GAGACACGCTAACGTTCACCAC	
ERG11	Forward	TCGGTCCATCTCTGTTTCTT	[31]
	Reverse	GAACACTGGGGTGGTCAAGT	
ERG 3	Forward	AAGCGTGTGAACAAGGAC	[30]
	Reverse	GCGTAGGTCTTCTCTGTGA	
CgSNQ2	Forward	CGTCCTATGTCTTCCTTACACCATT	[30]
	Reverse	TTTGAACCGCTTTTGTCTCTGA	
CgURA3	Forward	GAAAACCAATCTTTGTGCTTCTCT	[32]
	Reverse	CATGAGTCTTAAGCAAGCAAATGT	

2.10. qPCR Experiment

Each of the sample assays with and without fluconazole exposure was performed with identical triple replicates on a StepOnePlus Real-Time PCR system using a PowerUp SYBR Green Master Mix kit (Applied Biosystems, Waltham, MA, USA). The specificity of the qPCR reaction products was evaluated by analyzing the dissociation curves (melting curve).

Each qPCR quantification reaction contained PowerUp SYBR Green Master Mix [2X], 0.8 μM of each primer pair, and 10 ng/μL of cDNA. The program consisted of an initial denaturation in the following 2 steps: 50 °C for 2 min and 95 °C for 10 min, followed by 40 cycles of denaturation at 95 °C for 15 s and extension at 60 °C for 1 min. In order to establish the cDNA dissociation temperature and identify the non-specific amplification reactions, the following parameters were used: 95 °C for 15 s, 60 °C for 1 min, and 95 °C for 15 s.

2.11. qPCR Reaction Analysis

The ΔCt comparative analysis method was used to establish the differences between the threshold cycles (Ct) of the resistance genes (*ERG11*, *ERG3*, *CgCDR1*, and *CgSNQ2*) and the endogenous reference control gene (*CgURA3*). The analysis of the results was performed using StepOnePlus Real-Time PCR™ software version 2.3 (Applied Biosystems, Waltham, MA, USA). This software also made the evaluation of the efficiency of each reaction possible.

2.12. Molecular Docking

The molecular docking procedure was carried out using the amino acid sequence and the predicted tertiary structures of the *ERG3*, *ERG11*, *CgCDR1*, and *CgSNQ2* proteins (Uniprot codes P50860, P50859, Q6FK23, and Q6FQN3, respectively) of *C. glabrata*. The maximum resolution of the available crystal structures corresponded to 2.0 Å. Protein models were downloaded from the online biological database CANDIDA-GENOME DATABASE, with the structural predictions coupled to the ALPHAFOLD program version 2.3.25 [33,34]. The FASTA format was considered preliminarily for the sequence handling for each protein. The nomenclature example for *ERG3* corresponded to the following: >CAGL0F01793g | ERG3 COORDS: ChrF_C_glabrata_CBS138: 201866-200772C, MDLVLETLDHYIFDDVYAKIAPVELQRGID
DSLVNALSLNKIVSNSTLLHETLSITNSLKRVNKDVYGLTPFLFDFTEKTYASLLPRNNLIREF
FSLWAVVTVFGLLLYLITASLSYVFVFDRTIFNHPKYLKNQMYLEIKLAVSAIPTMSLLTVPWF
MLELNGYSKLYYDVDWEHHGLRKLLIEYATFIFFTDCGIYLAHRWLHWPRVYKALHKPHH
KWLVCTPFASHAFHPVDGYFQSLSYHIYPMILPLHKISYLILFTFVNFWSVMIHDGQHMSN
NPVVNGTACHTVHHLYFNYNYGQFTTLWDRLGGSYRRPEDSLFDPKLKMDKKVLEKQA
RETAAYIQEVEGDDTDRVYNTDKKKTN.

InterProScan Analysis (Ref) confirmed the membership of the sequences with the protein families associated with the fluconazole resistance genes in *C. glabrata*. [35].

A structural and geometric suitability evaluation of the proteins of interest was executed using the SWISSMODEL-STRUCTURE ASSESSMENT (https://swissmodel.expasy.org/, accessed on 1 November 2023) platform. Metrics from Ramachandran plots (favorable angles) of > 96% and the MolProbity Score estimator of <1.0 were utilized [36].

The optimized formats .mol2 for fluconazole and .pdb for the three-dimensional protein structures of interest were used. The analysis generated by the AVOGADRO program version 1.2.0 was used for the energy optimization of the substrate (fluconazole) [37]. The exploratory docking process was developed using the open-source programs PyRx-AutoDock Vina version 0.8 and CHIMERA version 1.16, with a grid box size of 26 Å × 26 Å × 26 Å and the protein-loading and ligand files in .pdbqt extensions [38]. The molecular result metrics are presented with Delta-Gibbs values or in Gibbs free energy (ΔG kcal/mol).

2.13. Statistical Analysis

According to the normality tests of the relative expression results of each gene of interest, parametric (Student's t) or non-parametric (Wilcoxon rank) statistics were used to determine differences in the means or medians. An alpha (α) of 0.05 was established as the level of statistical significance.

The multivariate analysis was based on the Fisher's linear discriminant functions route to provide a predictive classification score against the susceptibility (resistant or susceptible and dose-dependent) of *C. glabrata* [39,40]. The variables included in the multivariate model were as follows: gene, strain, mean RQ without fluconazole, and mean RQ with fluconazole.

The discriminant analysis was chosen as the proposed predictive model for the investigative process based on the intrinsic discriminant (dependence) capability sought in the research as follows: the classification capacity of the fluconazole susceptibility from the numerical independent variables, where the mean RQs with and without fluconazole and where the categorical options were the gene and strain, the confirmatory capacities of the new cases added to the model, obtaining the linear combination of independent variables that maximized the distinction of a group's membership, the estimation or probability metric of classifying each particular case to a group of interest, and the absence of collinearity or multicollinearity among the independent variables [40].

This route suggested the generation of qualitative variables with numerical values associated with the categories (susceptible and dose-dependent = 1, resistant = 2; *ERG3* gene = 1, *CgCDR1* gene = 2, *CgSNQ2* gene = 3; strain 04-1A = 1, strain 07-1A = 2, strain 109-1A = 3, etc.). Numerical variables (the mean RQs with and without fluconazole) were assumed in their continuous values in the model. Statistical significance was considered with predictive model classification accuracies (discriminant function) of $\geq$ 70%. The biostatistical analysis was supported by the licensed IBM SPSS package version 26 [41].

3. Results

3.1. Analysis of the Relative Gene Expression

As a result of the fluconazole susceptibility tests, the 16 strains were classified as follows: 6 were resistant (samples 06-3A, 95-1A, 116-5137-1184-1A, and 65-3) and 10 were susceptible and dose-dependent (samples 11-3A, 04-1A, 16-5A, 24-5A, 25-1A, 20-2A, 14-5A, 07-1A, 68-4, and 109-1A). For the analysis of the relative expression results of the *CgCDR1*, *ERG11*, *ERG3*, and *CgSNQ2* genes, the value of the relative expression of the *CgURA3* gene, corresponding to 1, was used as a reference. In other words, relative expression results greater than 1 were considered to be overexpressed. Additionally, through parametric and non-parametric analysis routes, the medians and/or averages of the relative expression levels of genes based on exposure or non-exposure to fluconazole were compared to determine the differences in the expression levels.

As observed in Figure 1, the relative expression (RQ) levels of the studied genes in the 16 strains, under the condition of exposure and non-exposure to fluconazole, regardless of the antifungal susceptibility profile, indicated that the relative expression levels of the exposed strains were higher than those of the non-exposed strains in *ERG3* (the difference in the RQs between the exposed and non-exposed strains was 1.61, with $p = 0.776$), followed by *CgSNQ2* (the difference in the RQs between the exposed strains and the non-exposed strains was 1, with $p = 0.044$), and, lastly, *CgCDR1* (the difference in the RQs between the exposed strains and the non-exposed strains was 0.3, with $p = 0.523$). In contrast, the *ERG11* gene in the strains exposed to fluconazole showed higher relative expression levels than the non-exposed strains (the difference in the RQs between the exposed strains and the non-exposed strains was -0.38, with $p = 0.034$).

Of the six strains classified as resistant to fluconazole, regardless of exposure to the antifungal, all overexpressed at least one gene, and it is noteworthy that the *ERG3* gene was overexpressed in all of them, the *CgCDR1* gene was overexpressed in four (06-3A, 184-1A, 116-5, and 65-3), the *ERG11* gene was overexpressed in two (06-3A and 65-3), and the *CgSNQ2* gene was overexpressed in two (95-1A and 65-3).

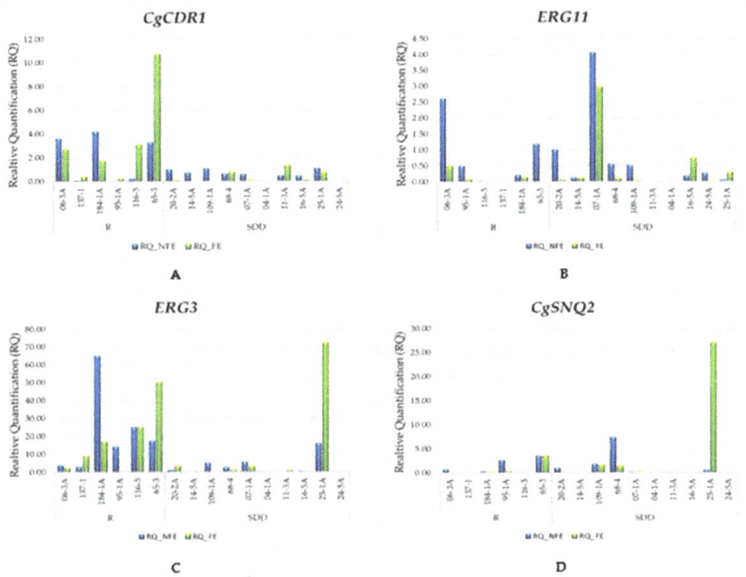

Figure 1. (**A**) Relative expression levels of the *CgCDR1* gene for strains not exposed vs. strains exposed to fluconazole. (**B**) Relative expression levels of the *ERG11* gene for strains not exposed vs. strains exposed to fluconazole. (**C**) Relative expression levels of the *ERG3* gene for strains not exposed vs. strains exposed to fluconazole. (**D**) Relative expression levels of the *CgSNQ2* gene for strains not exposed vs. strains exposed to fluconazole.

Strain 65-3 had higher relative expression of the *CgCDR1* and *ERG3* genes compared to the other strains, contrasting with a very low relative expression of *ERG11*. Additionally, under the condition of no exposure to fluconazole, it overexpressed all four genes studied, while the same strain exposed to the antifungal did not overexpress the *ERG11* gene. A similar phenomenon occurred with strain 06-3A, which, without being exposed, overexpressed the *CgCDR1*, *ERG11*, and *ERG3* genes, and when exposed to the antifungal, it did not overexpress the *ERG11* gene.

The descriptive analysis and the results of the parametric and non-parametric tests on the relative expression levels of the genes studied are reported in Table 3.

As shown in Table 1 and Figure 2, the relative expression levels of the *CgCDR1* gene in the fluconazole-exposed, resistant strains were higher than those in the non-exposed ones. However, these differences were not statistically significant (exposed, 2.22; non-exposed, 1.76; $p = 0.231$). On the contrary, the fluconazole-exposed, susceptible, dose-dependent strains were found to have lower relative expression levels than the non-exposed ones, and a statistically significant difference (RQ value for the susceptible, dose-dependent strains that were exposed was 0.11, and for the non-exposed, it was 0.59, with $p = 0.046$) was identified. According to the classification based on the susceptibility profile, higher RQ medians were obtained for the resistant strains, both for those exposed and those not exposed to fluconazole, compared to the susceptible, dose-dependent ones.

Table 3. Relative expression values of genes *CgCDR1*, *ERG11*, *ERG3*, and *CgSNQ2* according to susceptibility profile and exposure to the antifungal.

Analyzed Gene	Exposure to the Antifungal	Resistant One-Tailed Analysis						Susceptible and Dose-Dependent One-Tailed Analysis					
		Mean	Median	Minimum	Maximum	DE	*p*	Mean	Median	Minimum	Maximum	DE	*p*
CgCDR1	Without Fluconazole	1.91	1.76	0.05	4.22	1.99	0.231 [1]	0.6	0.59	0.04	1.09	0.38	0.046 [1]
	With Fluconazole	3.15	2.22	0.24	10.75	3.89		0.33	0.11	0.01	1.33	0.45	
ERG11	Without fluconazole	0.75	0.35	0.01	2.6	1.00	0.065 [2]	0.67	0.22	0	4.06	1.23	0.086 [2]
	With fluconazole	0.12	0.04	0.01	0.49	0.18		0.43	0.08	0	2.97	0.91	
ERG3	Without fluconazole	21.39	15.77	2.93	65	22.9	0.343 [1]	3.09	0.71	0	15.9	4.97	0.399 [1]
	With fluconazole	17.28	12.94	0.01	50.3	18.64		8.1	0.66	0.01	72.12	22.51	
CgSNQ2	Without fluconazole	1.22	0.52	0.01	3.59	1.53	0.014 [1]	1.1	0.12	0	7.4	2.29	0.166 [1]
	With fluconazole	0.69	0.11	0	3.59	1.42		3.03	0.05	0	26.95	8.42	

[1], One-tailed Wilcoxon rank-sum test; [2], one-tailed Student's *t*-test.

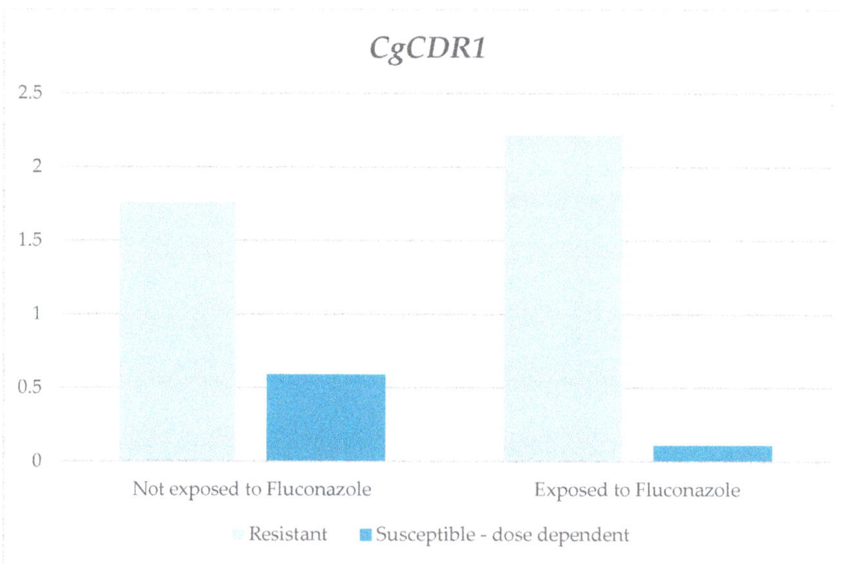

Figure 2. Relative expression of the *CgCDR1* gene according to susceptibility profile and the exposure to fluconazole.

The relative expression levels of the *ERG11* gene in the resistant and susceptible, dose-dependent strains were lower in those exposed to fluconazole compared to those not exposed. Nevertheless, these differences were not significant (RQ-resistant for those exposed, 0.12, and for those not exposed, 0.75, with $p = 0.065$; RQ-susceptible and dose-dependent for those exposed, 0.43, and for those not exposed, 0.67, with $p = 0.086$). Between the two groups and according to the susceptibility profiles, there were higher average RQs in the susceptible, dose dependent strains than in the resistant ones (Table 1 and Figure 3).

On the other hand, the relative expression levels of the *ERG3* gene in the resistant and susceptible, dose-dependent strains exposed to fluconazole were lower than in those not exposed, with statistically non-significant differences (RQ-resistant for those exposed, 12.9, and for those not exposed, 15.77, with $p = 0.343$; RQ-susceptible and dose-dependent for those exposed, 0.66, and for those not exposed, 0.71, with $p = 0.399$). There were higher RQ medians in the resistant strains, for both those exposed and those not exposed to fluconazole, than in the susceptible, dose-dependent strains, as can be observed in Table 1 and Figure 4.

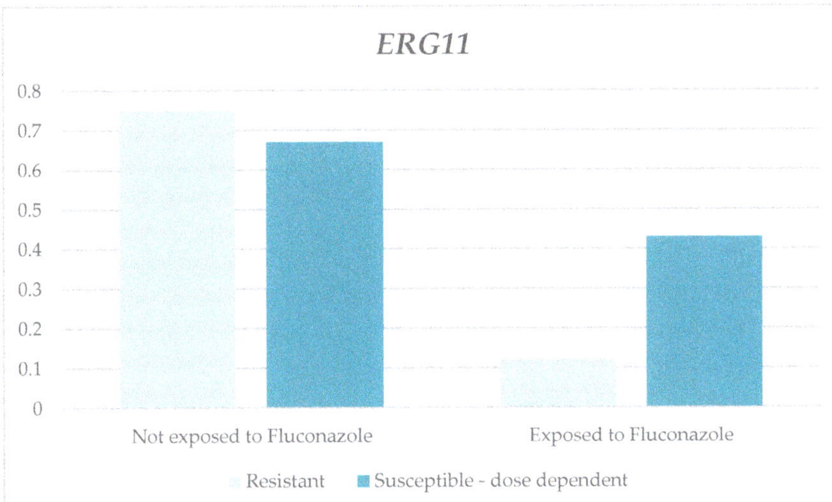

Figure 3. Relative expression of the *ERG11* gene according to susceptibility profile and the exposure to fluconazole.

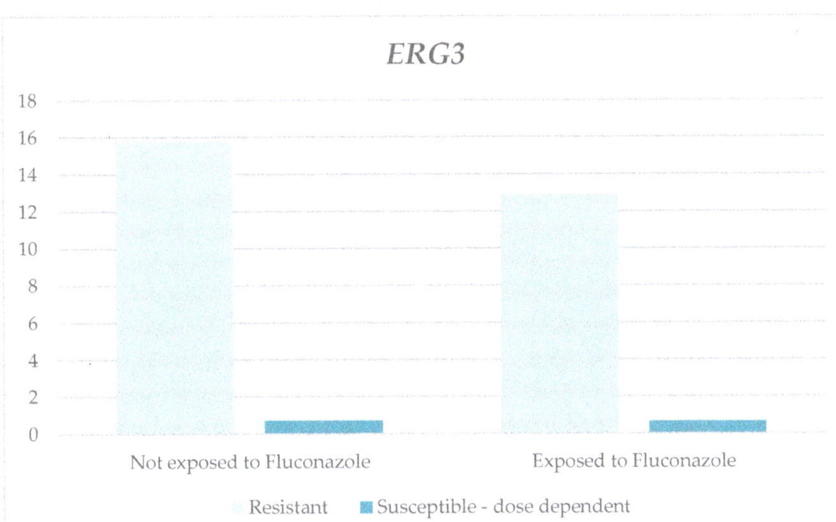

Figure 4. Relative expression of the *ERG3* gene according to susceptibility profile and the exposure to fluconazole.

Lastly, the relative expression levels of the *CgSNQ2* gene in both the resistant and susceptible, dose-dependent strains were lower in the strains exposed to the antifungal.

However, the differences were only significant for the resistant group (RQ-resistant for those exposed, 0.11, and for those not exposed, 0.52, with $p = 0.014$; RQ-susceptible and dose-dependent for those exposed, 0.05, and for those not exposed, 0.12, with $p = 0.166$). The RQ medians were higher in both the exposed and not-exposed resistant strains compared to the susceptible, dose-dependent strains (Table 1 and Figure 5).

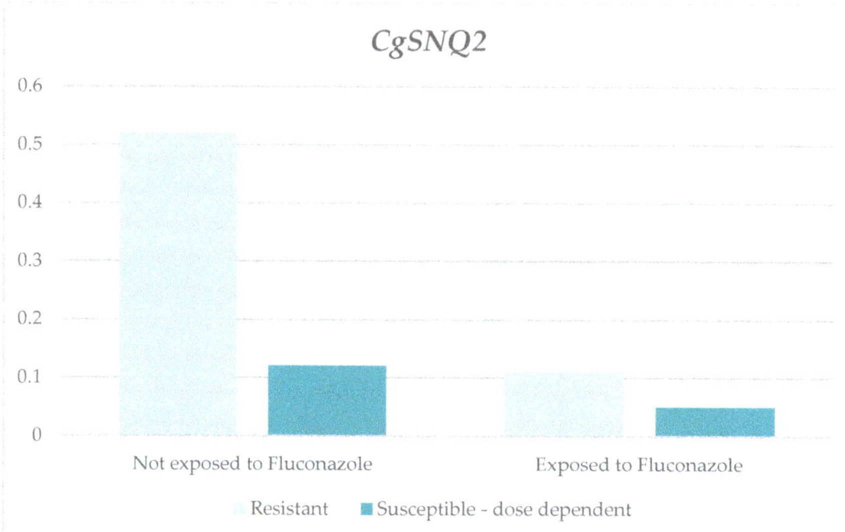

Figure 5. Relative expression of the *CgSNQ2* gene according to susceptibility profile and the exposure to fluconazole.

3.2. Results of the Predictive Model on the Susceptibility Profile

The analysis model included four categorical and continuous predictor variables (gene, strain, RQ mean without fluconazole, and RQ mean with fluconazole) and one categorical predictor variable (fungus susceptibility profile). The multivariate evaluation through the route of the Fisher's linear discriminant function or a dichotomous prediction (sum of the estimation products) generated the coefficients for the classification of susceptibility/susceptibility profile, as shown in Table 4.

Table 4. Coefficients in the classification function for the susceptible dose-dependent (SDD) or resistant profiles.

	General Susceptibility Profile	
Variables in the Model	**Resistant**	**SDD**
Gene analyzed	4.141	3.730
Categorized strain	0.369	0.276
RQ mean without fluconazole	0.363	0.302
RQ mean with fluconazole	0.111	0.042
Constant	−8.009	−5.820

The overall comparative classification results between the original susceptibility profile findings by the macrodilution tests as compared to the classifications predicted by the model yielded a classification accuracy of 73.5% (>0.7).

The utilization of the coefficient results is described below in Table 5. The findings in the research corresponded to the *CgCDR1* gene (categorical value = 2) from *C. glabrata* strain 07-1A (categorical value = 2), with an RQ mean without fluconazole of 0.57 and an RQ mean with fluconazole of 0.22. They were replaced with the product of the coefficients, and the products were added, as shown below.

Table 5. Use of the results of the model coefficients.

	Susceptibility Profile		Forecasted Profile	
Variables in the Model	**Resistant**	**SDD**	**Resistant**	**SDD**
Gene analyzed	4.141	3.730	2 × 4.141	2 × 3.730
Categorized strain	0.369	0.276	2 × 0.369	2 × 0.276
RQ mean without fluconazole	0.363	0.302	0.57 × 0.363	0.57 × 0.302
RQ mean with fluconazole	0.111	0.042	0.22 × 0.111	0.22 × 0.042
Constant	−8.009	−5.820	1 × (−8.009)	1 × (−5.820)
		Total	1.2425	2.3738

The total values were compared, and the highest value defined the forecasted profile. In this case, under the example's characteristics, the SDD category was forecasted. This classification was considered correct (accurate) by the model in 73.5% of the cases.

3.3. Molecular Docking

The molecular processing using basic structural bioinformatic techniques (rigid docking) made obtaining the ΔG values for the relationship between the proteins expressed by the *C. glabrata* genes *ERG3*, *ERG11*, *CgCDR1*, and *CgSNQ2* and the fluconazole substrate possible. Table 6 describes the findings of the energetic molecular metrics for the main positions (interaction sites) of fluconazole and the mentioned proteins.

Table 6. Representation of the ΔG (kcal/mol) values for the various positions generated by the docking model for fluconazole and the proteins of interest in *C. glabrata*.

		Fluconazole	
Protein	**Model**	**ΔG (Kcal/mol)**	**Number of Contacts**
ERG3	1.	−6.2	19
	2.	−6.1	13
	3.	−6.0	6
	4.	−5.9	11
	5.	−5.8	6
ERG11	1.	−2.8	12
	2.	−2.6	11
	3.	−2.5	11
	4.	−2.4	10
	5.	−2.2	9
CgCDR1	1.	−7.0	14
	2.	−6.4	14
	3.	−6.4	15
	4.	−6.2	14
	5.	−6.2	11
CgSNQ2	1.	−7.6	17
	2.	−7.5	12
	3.	−7.4	12
	4.	−7.2	11
	5.	−7.1	11

The docking model also revealed the relationships between the constituent amino acids and fluconazole at the docking sites. Figure 6 shows two representative models of the process for fluconazole and *ERG11* and fluconazole and *CgCDR1*.

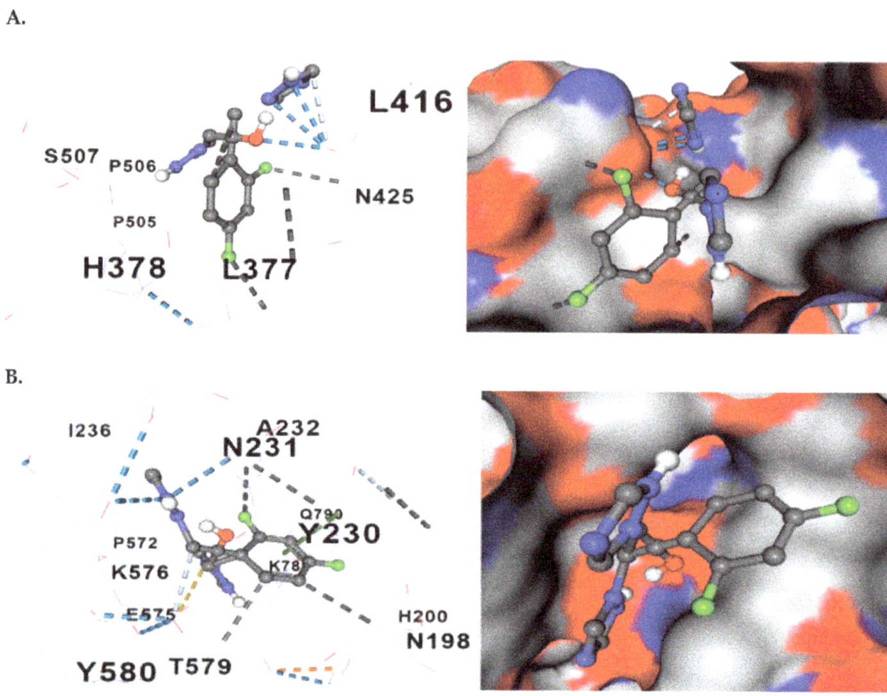

Figure 6. Comparative representations of fluconazole and the proteins of interest docking. (**A**) Three-dimensional presentation of the molecular relationship (docking) between fluconazole and *ERG11*. (**B**) Three-dimensional presentation of the molecular relationship between fluconazole and *CgCDR1*. Hydrophobic interactions with the residues are shown with dashed gray lines and with gray surfaces on the proteins.

The results of the docking revealed a median of 10 detected contacts for the interaction between fluconazole and *ERG11*, and for the interactions between fluconazole and *CgCDR1*, *ERG3, and CgSNQ2*, there was a median of 13.

4. Discussion

This study addressed the quantification of gene expression in the species *C. glabrata*, encoding either the drug efflux transporters or the enzymes involved in the ergosterol biosynthesis pathway.

The genes encoding the addressed efflux transporters were *CgCDR1* and *CgSNQ2*, and their expression findings allowed us to identify that regardless of the fluconazole susceptibility profile, the strains exposed to the antifungal had higher relative expression levels than those not exposed, indicating a greater amount of pharmacological efflux transporters available for reducing the intracellular fluconazole concentrations in the experimental group treated with the antifungal. Moreover, it was possible to identify that the resistant strains also had higher expression levels than those classified as susceptible and dose-dependent.

Complementary findings from the molecular perspective showed that the detected phenotype of the *C. glabrata* activity established that there was a limited pharmacological response in the scenario of *CgCDR1* overexpression and that the affinity between fluconazole and *CgSNQ2* was superior to those of the other genes of interest in this study. Although no other studies using computational techniques for molecular docking were found, similar results were found in other studies where resistance to azoles was related to the overexpression of genes encoding efflux pumps, such as the *CgCDR1* and *CgSNQ2* genes related to the ATP-Binding Cassette (ABC) superfamily of proteins [11,14,23,42–47].

Researchers Karl, Joseph, and Tomas conducted a study on resistance in *C. albicans* and identified that the deletion of the *ERG11* gene was associated with the positive regulation of *ERG3* expression, as well as the presence of enzyme inhibitor drugs and genetic alterations in ergosterol biosynthesis, leading to increases in *ERG3* mRNA levels [48]. Although this finding does not correspond to the same species we used, it may behave biologically similarly to the *Candida* genus, since our study identified generally very low *ERG11* expression values, most of them even lower than one, which corresponded to the expression of the reference strain. In contrast, much higher *ERG3* expression values were evidenced.

Moreover, our results showed the overexpression of *ERG3* in almost all strains (both resistant and susceptible, dose-dependent), a situation that could be extrapolated to a lower susceptibility to the action of fluconazole. Nevertheless, *ERG3* expression stood out for its very high values in those classified as resistant compared to those classified as susceptible and dose-dependent. It is plausible that *ERG3* overexpression is an early cellular stress response to try to maintain membrane integrity in the presence of a blockade generated by azoles on an enzyme encoded by *ERG11*, particularly in *C. glabrata*. Being intrinsically less susceptible to fluconazole, this was a finding consistent with the results of molecular docking that would indicate a limited affinity between fluconazole and *ERG11* [49–51].

The available studies have focused on investigating loss-of-function mutations in *ERG3*, which may be biological responses to overexpression in which the inactivity of the enzyme encoded by *ERG3* resulting from the loss-of-function mutations leads to decreases in the synthesis of toxic intermediate sterols in the presence of azoles, favoring the survival of the fungus [14,18,20]. Nevertheless, these studies have not focused on quantifying expression, and so references that could provide valuable information about *ERG3* expression and azole/fluconazole resistance in *Candida*, especially in *C. glabrata*, are scarce, and the relationship between *ERG3* expression and azole resistance is not clear.

When the strains resistant to the antifungal were exposed, the expression of the *ERG11* gene decreased. The relative expression of the *ERG11* gene was low in both the resistant and susceptible, dose-dependent strains, as well as in the strains exposed to the antifungal compared to those not exposed, and only three strains showed expression levels above one. These results for *ERG11* gene expression may be related, as one of the biological conditions that explain the lower intrinsic susceptibility of *C. glabrata* to fluconazole compared to other *Candida* species could indicate that low gene expression may affect lanosterol 14-alpha-demethylase synthesis and decrease pharmacological efficacy by reducing the size of the pharmacological target [52].

There is limited scientific evidence on this subject, and the studies found do not report specific *ERG11* expression values in *C. glabrata*, making it necessary to propose more research to expand the conceptual framework.

This study presents several limitations. Firstly, 1% agarose gel electrophoresis was used to evaluate the RNA integrity, and this is a methodology considered to be limited in terms of identifying susceptibility and precision. It is important to note that there are currently more advanced and sensitive methodologies, such as the bioanalyzer, which allow for a more exhaustive and quantitative evaluation of RNA quality. Nonetheless, this technology could not be used due to a lack of technological availability in the region. Additionally, the scope of the study did not allow for the evaluation of other molecular mechanisms, such as point mutations in the genes involved in ergosterol synthesis, due to financial constraints.

Despite the mentioned restrictions, the findings obtained provide a preliminary analysis of the probable role played by exposure to fluconazole in the development of resistance in *C. glabrata* isolates. These results support the recommendation to avoid unnecessary and inadequate prescription of this antifungal, as well as its prolonged use, and they emphasize the importance of basing its administration on accurate diagnosis confirmation and species susceptibility profiling.

5. Conclusions

The phenomenon of resistance is complex and depends on many factors inherent to this fungus, its host, and the antifungal drug used. It was possible to identify that the strains resistant to fluconazole had significant expression levels of genes encoding drug efflux pumps belonging to the ATP-Binding Cassette superfamily, and there was a much higher expression of the *ERG3* gene compared to the other genes involved in the analyses. This may be a particular characteristic of the *C. glabrata* species, with its comparatively low expression of the *ERG11* gene. These findings should motivate further research that favors a comprehensive understanding of the adaptation–resistance dynamics to antimicrobials.

To date, no similar work has attempted to understand this resistance phenomenon through the integration of categorical and numerical variables derived from molecular observations and fluconazole responses in *C. glabrata*. Therefore, this contribution may be important for the scientific community, and it may be taken as a reference for studies involving complementary routes that integrate the variables in predictive models that complement and streamline such analyses. In addition, this type of molecular analysis allows for integrating an explanatory scenario of the relationship between widely used antimicrobials and the biological activity represented by *C. glabrata*. Considering that such studies can help scientific and clinical communities better understand the biological basis of resistance, they may contribute key elements to solving this problem with predictive, applicative, and therapeutic precision.

Supplementary Materials: The following supporting information can be downloaded at https://www.mdpi.com/article/10.3390/jof10070509/s1. Raw data: *Candida glabrata* Database.

Author Contributions: L.Y.C.P., conceptualization, software, formal analysis, writing—original draft, project administration, funding acquisition, preparation, writing—review and editing, visualization; A.E.R.R., conceptualization, validation, investigation, data curation, writing—review and editing, project administration, and funding acquisition; J.E.P.C., methodology, supervision, project administration, funding acquisition; J.M.P.-A., conceptualization, software, formal analysis, writing—review and editing, visualization. All authors have read and agreed to the published version of the manuscript.

Funding: This research was funded by the Universidad Católica de Manizales: Acuerdo No 036 and the Universidad de Caldas: Código 0585120.

Institutional Review Board Statement: The study was conducted in accordance with the Declaration of Helsinki and approved by the Ethics Committees of the Universidad Católica de Manizales and the Universidad de Caldas.

Informed Consent Statement: This study did not involve human subjects, nor did it use sensitive information from medical records or other data, therefore informed consent was not required.

Data Availability Statement: The original contributions presented in the study are included in the article/Supplementary Materials, further inquiries can be directed to the corresponding author.

Acknowledgments: We extend our sincere gratitude to all the institutions and staff who made this research possible.

Conflicts of Interest: The authors declare no conflict of interest.

References

1. Whaley, S.G.; Berkow, E.L.; Rybak, J.M.; Nishimoto, A.T.; Barker, K.S.; Rogers, P.D. Azole Antifungal Resistance in *Candida albicans* and Emerging Non-albicans *Candida* Species. *Front. Microbiol.* **2016**, *7*, 2173. [CrossRef] [PubMed]
2. Arendrup, M.C.; Patterson, T.F. Multidrug-Resistant *Candida*: Epidemiology, Molecular Mechanisms, and Treatment. *J. Infect. Dis.* **2017**, *216* (Suppl. S3), S445–S451. [CrossRef] [PubMed]

3. Zuluaga Rodríguez, A.; de Bedout Gómez, C.; Agudelo Restrepo, C.A.; Hurtado Parra, H.; Arango Arteaga, M.; Restrepo Moreno, Á.; González Marín, Á. Sensibilidad a fluconazol y voriconazol de especies de *Candida* aisladas de pacientes provenientes de unidades de cuidados intensivos en Medellín, Colombia (2001–2007). *Rev. Iberoam. Micol.* **2010**, *27*, 125–129. [CrossRef] [PubMed]
4. Cortés, J.A.; Reyes, P.; Gómez, C.H.; Cuervo, S.I.; Rivas, P.; Casas, C.A.; Sánchez, R. Clinical and epidemiological characteristics and risk factors for mortality in patients with candidemia in hospitals from Bogotá, Colombia. *Braz. J. Infect. Dis.* **2014**, *18*, 631–637. [CrossRef] [PubMed]
5. Maldonado, N.A.; Cano, L.E.; De Bedout, C.; Arbeláez, C.A.; Roncancio, G.; Tabares, M.; Robledo, C.G.; Robledo, J. Association of clinical and demographic factors in invasive candidiasis caused by fluconazole-resistant *Candida* species: A study in 15 hospitals, Medellín, Colombia 2010–2011. *Diagn. Microbiol. Infect. Dis.* **2014**, *79*, 280–286. [CrossRef] [PubMed]
6. Cortés, J.A.; Ruiz, J.F.; Melgarejo-Moreno, L.N.; Lemos, E.V. Candidemia en colombia. *Biomédica* **2020**, *40*, 195–207. [CrossRef]
7. Vides-Peña, C.A.; Bolaño-Ardila, N.D.; Vides-Peña, M.V.; Córdoba, S.B. Epidemiología de la candidemia y sensibilidad in vitro frente a azoles, anfotericina B y caspofungina en una institución de salud de Valledupar, Colombia. *Med. Lab.* **2015**, *21*, 255–266. [CrossRef]
8. Yao, D.; Chen, J.; Chen, W.; Li, Z.; Hu, X. Mechanisms of azole resistance in clinical isolates of *Candida glabrata* from two hospitals in China. *Infect. Drug Resist.* **2019**, *12*, 771–781. [CrossRef]
9. Shahrokhi, S.; Noorbakhsh, F.; Rezaie, S. Quantification of CDR1 Gene Expression in Fluconazole Resistant Candida Glabrata Strains Using Real-time PCR. *Iran. J. Public Health* **2017**, *46*, 1118–1122. [PubMed]
10. Tortorano, A.M.; Prigitano, A.; Morroni, G.; Brescini, L.; Barchiesi, F. Candidemia: Evolution of Drug Resistance and Novel Therapeutic Approaches. *Infect. Drug Resist.* **2021**, *14*, 5543–5553. [CrossRef]
11. Whaley, S.G.; Zhang, Q.; Caudle, K.E.; Rogers, P.D. Relative Contribution of the ABC Transporters Cdr1, Pdh1, and Snq2 to Azole Resistance in *Candida glabrata*. *Antimicrob. Agents Chemother.* **2018**, *62*, e01070-18. [CrossRef] [PubMed]
12. Maheronnaghsh, M.; Teimoori, A.; Dehghan, P.; Fatahinia, M. The evaluation of the overexpression of the *ERG-11*, *MDR-1*, *CDR-1*, and *CDR-2* genes in fluconazole-resistant *Candida albicans* isolated from Ahvazian cancer patients with oral candidiasis. *J. Clin. Lab. Anal.* **2022**, *36*, e24208. [CrossRef] [PubMed]
13. Paul, S.; Singh, S.; Sharma, D.; Chakrabarti, A.; Rudramurthy, S.M.; Ghosh, A.K. Dynamics of in vitro development of azole resistance in *Candida tropicalis*. *J. Glob. Antimicrob. Resist.* **2020**, *22*, 553–561. [CrossRef] [PubMed]
14. Sakagami, T.; Kawano, T.; Yamashita, K.; Yamada, E.; Fujino, N.; Kaeriyama, M.; Fukuda, Y.; Nomura, N.; Mitsuyama, J.; Suematsu, H.; et al. Antifungal susceptibility trend and analysis of resistance mechanism for *Candida* species isolated from bloodstream at a Japanese university hospital. *J. Infect. Chemother.* **2019**, *25*, 34–40. [CrossRef] [PubMed]
15. Healey, K.R.; Perlin, D.S. Fungal Resistance to Echinocandins and the MDR Phenomenon in *Candida glabrata*. *J. Fungi* **2018**, *4*, 105. [CrossRef]
16. Zare-Bidaki, M.; Maleki, A.; Ghanbarzadeh, N.; Nikoomanesh, F. Expression pattern of drug-resistance genes *ERG11* and *TAC1* in *Candida albicans* Clinical isolates. *Mol. Biol. Rep.* **2022**, *49*, 11625–11633. [CrossRef]
17. Liu, Z.; Myers, L.C. Mediator Tail Module Is Required for Tac1-Activated *CDR1* Expression and Azole Resistance in *Candida albicans*. *Antimicrob. Agents Chemother.* **2017**, *61*, e01342-17. [CrossRef]
18. Feng, W.; Yang, J.; Xi, Z.; Ji, Y.; Zhu, X.; Yang, L.; Ma, Y. Regulatory Role of ERG3 and Efg1 in Azoles-Resistant Strains of *Candida albicans* Isolated from Patients Diagnosed with Vulvovaginal Candidiasis. *Indian J. Microbiol.* **2019**, *59*, 514–524. [CrossRef] [PubMed]
19. Liu, Z.; Myers, L.C. *Candida albicans* Swi/Snf and Mediator Complexes Differentially Regulate Mrr1-Induced *MDR1* Expression and Fluconazole Resistance. *Antimicrob. Agents Chemother.* **2017**, *61*, e01344-17. [CrossRef]
20. Vu, B.G.; Thomas, G.H.; Moye-Rowley, W.S. Evidence that Ergosterol Biosynthesis Modulates Activity of the Pdr1 Transcription Factor in *Candida glabrata*. *mBio* **2019**, *10*, 1–20. [CrossRef]
21. Sanguinetti, M.; Posteraro, B.; Fiori, B.; Ranno, S.; Torelli, R.; Fadda, G. Mechanisms of azole resistance in clinical isolates of *Candida glabrata* collected during a hospital survey of antifungal resistance. *Antimicrob. Agents Chemother.* **2005**, *49*, 668–679. [CrossRef] [PubMed]
22. El Said, M.; Badawi, H.; Gamal, D.; Salem, D.; Dahroug, H.; El-Far, A. Detection of ERG11 gene in fluconazole resistant urinary *candida* isolates. *Egypt. J. Immunol.* **2022**, *29*, 134–147. [CrossRef] [PubMed]
23. Spettel, K.; Barousch, W.; Makristathis, A.; Zeller, I.; Nehr, M.; Selitsch, B.; Lackner, M.; Rath, P.-M.; Steinmann, J.; Willinger, B. Analysis of antifungal resistance genes in *Candida albicans* and *Candida glabrata* using next generation sequencing. *PLoS ONE* **2019**, *14*, e0210397. [CrossRef] [PubMed]
24. Luna-Tapia, A.; Willems, H.M.E.; Parker, J.E.; Tournu, H.; Barker, K.S.; Nishimoto, A.T.; Rogers, P.D.; Kelly, S.L.; Peters, B.M.; Palmer, G.E. Loss of Upc2p-Inducible *ERG3* Transcription Is Sufficient to Confer Niche-Specific Azole Resistance without Compromising *Candida albicans* Pathogenicity. *mBio* **2018**, *9*, 1–17. [CrossRef] [PubMed]
25. Robbins, N.; Cowen, L.E. Antifungal drug resistance: Deciphering the mechanisms governing multidrug resistance in the fungal pathogen *Candida glabrata*. *Curr. Biol.* **2021**, *31*, R1520–R1523. [CrossRef] [PubMed]
26. Vu, B.G.; Moye-Rowley, W.S. Azole-Resistant Alleles of *ERG11* in *Candida glabrata* Trigger Activation of the Pdr1 and Upc2A Transcription Factors. *Antimicrob. Agents Chemother.* **2022**, *66*, e0209821. [CrossRef] [PubMed]

27. Clinical and Laboratory Standards Institute (CLSI). *Reference Method for Broth Dilution Antifungal Susceptibility Testing of Yeasts*, 4th ed.; M27-A4; CLSI: Wayne, PA, USA, 2017.
28. Clinical and Laboratory Standards Institute (CLSI). *Performance Standards for Antifungal Susceptibility Testing of Yeasts M60*; CLSI: Wayne, PA, USA, 2017.
29. Méndez, C.C.; Sánchez, E.G.; Martín-Mazuelos, E. Actualización de los métodos de estudio de sensibilidad in vitro a los antifúngicos. *Enfermedades Infecc. Microbiol. Clin.* **2019**, *37*, 32–39. [CrossRef] [PubMed]
30. Caudle, K.E.; Barker, K.S.; Wiederhold, N.P.; Xu, L.; Homayouni, R.; Rogers, P.D. Genomewide expression profile analysis of the *Candida glabrata* Pdr1 regulon. *Eukaryot. Cell* **2011**, *10*, 373–383. [CrossRef]
31. Silva, D.B.d.S.; Rodrigues, L.M.C.; de Almeida, A.A.; de Oliveira, K.M.P.; Grisolia, A.B. Novel point mutations in the *ERG11* gene in clinical isolates of azole resistant *Candida* species. *Mem. Inst. Oswaldo Cruz* **2016**, *111*, 192–199. [CrossRef] [PubMed]
32. Szweda, P.; Gucwa, K.; Romanowska, E.; Dzierżanowska-Fangrat, K.; Naumiuk, Ł.; Brillowska-Dąbrowska, A.; Wojciechowska-Koszko, I.; Milewski, S. Mechanisms of azole resistance among clinical isolates of *Candida glabrata* in Poland. *J. Med. Microbiol.* **2015**, *64*, 610–619. [CrossRef] [PubMed]
33. Skrzypek, M.S.; Binkley, J.; Binkley, G.; Miyasato, S.R.; Simison, M.; Sherlock, G. The *Candida* Genome Database (CGD): Incorporation of Assembly 22, systematic identifiers and visualization of high throughput sequencing data. *Nucleic Acids Res.* **2017**, *45*, D592–D596. [CrossRef] [PubMed]
34. Jumper, J.; Evans, R.; Pritzel, A.; Green, T.; Figurnov, M.; Ronneberger, O.; Tunyasuvunakool, K.; Bates, R.; Žídek, A.; Potapenko, A.; et al. Highly accurate protein structure prediction with AlphaFold. *Nature* **2021**, *596*, 583–589. [CrossRef] [PubMed]
35. Paysan-Lafosse, T.; Blum, M.; Chuguransky, S.; Grego, T.; Pinto, B.L.; A Salazar, G.; Bileschi, M.L.; Bork, P.; Bridge, A.; Colwell, L.; et al. InterPro in 2022. *Nucleic Acids Res.* **2023**, *51*, D418–D427. [CrossRef] [PubMed]
36. Waterhouse, A.; Bertoni, M.; Bienert, S.; Studer, G.; Tauriello, G.; Gumienny, R.; Heer, F.T.; De Beer, T.A.P.; Rempfer, C.; Bordoli, L.; et al. SWISS-MODEL: Homology modelling of protein structures and complexes. *Nucleic Acids Res.* **2018**, *46*, W296–W303. [CrossRef] [PubMed]
37. Hanwell, M.D.; Curtis, D.E.; Lonie, D.C.; Vandermeersch, T.; Zurek, E.; Hutchison, G.R. Avogadro: An advanced semantic chemical editor, visualization, and analysis platform. *J. Cheminform.* **2012**, *4*, 17. [CrossRef] [PubMed]
38. Pettersen, E.F.; Goddard, T.D.; Huang, C.C.; Couch, G.S.; Greenblatt, D.M.; Meng, E.C.; Ferrin, T.E. UCSF Chimera?A visualization system for exploratory research and analysis. *J. Comput. Chem.* **2004**, *25*, 1605–1612. [CrossRef] [PubMed]
39. Cheng, W.-C.; Chen, L.-H.; Jiang, C.-R.; Deng, Y.-M.; Wang, D.-W.; Lin, C.-H.; Jou, R.; Wang, J.-K.; Wang, Y.-L. Sensible Functional Linear Discriminant Analysis Effectively Discriminates Enhanced Raman Spectra of *Mycobacterium* Species. *Anal. Chem.* **2021**, *93*, 2785–2792. [CrossRef] [PubMed]
40. Safo, S.E.; Min, E.J.; Haine, L. Sparse linear discriminant analysis for multiview structured data. *Biometrics* **2022**, *78*, 612–623. [CrossRef] [PubMed]
41. IBM SPSS Software [Internet]. 2023. Available online: https://www.ibm.com/es-es/spss (accessed on 28 June 2023).
42. Castanheira, M.; Deshpande, L.M.; Davis, A.P.; Carvalhaes, C.G.; Pfaller, M.A. Azole resistance in *Candida glabrata* clinical isolates from global surveillance is associated with efflux overexpression. *J. Glob. Antimicrob. Resist.* **2022**, *29*, 371–377. [CrossRef] [PubMed]
43. Gohar, A.A.; Badali, H.; Shokohi, T.; Nabili, M.; Amirrajab, N.; Moazeni, M. Expression Patterns of ABC Transporter Genes in Fluconazole-Resistant *Candida glabrata*. *Mycopathologia* **2017**, *182*, 273–284. [CrossRef] [PubMed]
44. Won, E.J.; Choi, M.J.; Kim, M.-N.; Yong, D.; Lee, W.G.; Uh, Y.; Kim, T.S.; Byeon, S.A.; Lee, S.Y.; Kim, S.H.; et al. Fluconazole-Resistant *Candida glabrata* Bloodstream Isolates, South Korea, 2008–2018. *Emerg. Infect. Dis.* **2021**, *27*, 779–788. [CrossRef] [PubMed]
45. Khalifa, H.O.; Arai, T.; Majima, H.; Watanabe, A.; Kamei, K. Genetic Basis of Azole and Echinocandin Resistance in Clinical *Candida glabrata* in Japan. *Antimicrob. Agents Chemother.* **2020**, *64*, e00783-20. [CrossRef]
46. Biswas, C.; Marcelino, V.R.; Van Hal, S.; Halliday, C.; Martinez, E.; Wang, Q.; Kidd, S.; Kennedy, K.; Marriott, D.; Morrissey, C.O.; et al. Whole Genome Sequencing of Australian *Candida glabrata* Isolates Reveals Genetic Diversity and Novel Sequence Types. *Front. Microbiol.* **2018**, *9*, 2946. [CrossRef] [PubMed]
47. Godinho, C.P.; Dias, P.J.; Ponçot, E.; Sá-Correia, I. The Paralogous Genes *PDR18* and *SNQ2*, Encoding Multidrug Resistance ABC Transporters, Derive from a Recent Duplication Event, PDR18 Being Specific to the *Saccharomyces* Genus. *Front. Genet.* **2018**, *9*, 476. [CrossRef]
48. Henry, K.W.; Nickels, J.T.; Edlind, T.D. Upregulation of *ERG* Genes in *Candida* Species by Azoles and Other Sterol Biosynthesis Inhibitors. *Antimicrob. Agents Chemother.* **2000**, *44*, 2693–2700. [CrossRef] [PubMed]
49. Iyer, K.R.; Robbins, N.; Cowen, L.E. The role of *Candida albicans* stress response pathways in antifungal tolerance and resistance. *iScience* **2022**, *25*, 103953. [CrossRef] [PubMed]
50. Berman, J.; Krysan, D.J. Drug resistance and tolerance in fungi. *Nat. Rev. Microbiol.* **2020**, *18*, 319–331. [CrossRef] [PubMed]

51. Hill, J.A.; O'meara, T.R.; Cowen, L.E. Fitness Trade-Offs Associated with the Evolution of Resistance to Antifungal Drug Combinations. *Cell Rep.* **2015**, *10*, 809–819. [CrossRef] [PubMed]
52. Song, J.L.; Harry, J.B.; Eastman, R.T.; Oliver, B.G.; White, T.C. The *Candida albicans* lanosterol 14-α-demethylase (*ERG11*) gene promoter is maximally induced after prolonged growth with antifungal drugs. *Antimicrob. Agents Chemother.* **2004**, *48*, 1136–1144. [CrossRef] [PubMed]

MDPI AG
Grosspeteranlage 5
4052 Basel
Switzerland
Tel.: +41 61 683 77 34

Journal of Fungi Editorial Office
E-mail: jof@mdpi.com
www.mdpi.com/journal/jof

www.ingramcontent.com/pod-product-compliance
Lightning Source LLC
Chambersburg PA
CBHW041954260926
51850CB00019B/2730
* 9 7 8 3 7 2 5 8 7 3 4 6 3 *